William Bernhard Tegetmeier

Pheasants : their natural History and practical Management

William Bernhard Tegetmeier

Pheasants : their natural History and practical Management

ISBN/EAN: 9783337024642

Printed in Europe, USA, Canada, Australia, Japan

Cover: Foto ©berggeist007 / pixelio.de

More available books at **www.hansebooks.com**

PHEASANTS

THEIR

Natural History and Practical Management.

BY

W. B. TEGETMEIER
(Member of the British Ornithologists' Union),
AUTHOR OF "THE NATURAL HISTORY OF THE CRANES," "TABLE AND MARKET POULTRY," ETC., ETC.

THIRD EDITION, ENLARGED.

ILLUSTRATED FROM LIFE BY MESSRS. J. G. MILLAIS, T. W. WOOD, P. SMIT, AND F. W. FROHAWK, ETC.

LONDON:

HORACE COX,
"THE FIELD" OFFICE, BREAM'S BUILDINGS, E.C.

1897.

(All rights reserved.)

PREFACE.

A DETAILED ACCOUNT of the natural history, habits, food, and treatment of the various species of Pheasants had long been a desideratum; this book was projected with a view to supply the want, in a more complete and comprehensive form than had hitherto been attempted. The extremely favourable reception which the previous editions met with, not only from the reviewers, but also from the general public, showed that the demand for such information was not over-estimated, whilst the opinions expressed by many of our highest authorities have led me to believe that the endeavour to combine ornithological research with practical experience in the management of this group of birds was not unattended with success.

In the following work I have given the natural history and general practical management, not only of the pheasants strictly adapted for the covert, but also of the allied species, which are the best adapted to our aviaries.

The progress of scientific exploration is continually bringing to light species of pheasants hitherto unknown;

some of these are well suited to our coverts, whilst others are regarded as ornamental birds. A few years since the only pheasant bred wild in England was the common species (*Phasianus colchicus*); our coverts now possess the Chinese (*P. torquatus*) and the Japanese (*P. versicolor*) species; whilst the Reeves's pheasant (*P. reevesii*), still more beautiful, and equally well adapted both for sporting and culinary purposes, has been recently introduced. In the same manner, our aviaries have recently been enriched by the addition of the Amherst pheasant (*Thaumalea amherstiæ*) and others, which, by their exquisite beauty, eclipse even the gorgeous coloration and elegant markings of the comparatively well-known Gold and Silver pheasants.

To indicate and illustrate these various species, to give as far as is known their natural history, to describe the best methods of rearing them in preserves and inclosed pheasantries, to enter into the numerous details respecting their food, management, protection, rearing, diseases, &c., is the object at which I have aimed in the preparation of this work.

In the following chapters I first treat of the natural history of the pheasants generally — their food, habits, nesting, &c. Then follows the consideration of their management in preserves, the details of the different methods of feeding the birds, their protection from their numerous enemies, the formation of coverts, &c. This is succeeded by an account of their treatment in inclosed pheasantries, the hatching of the eggs, rearing and feeding the young birds, and the prevention and cure of their diseases.

A detailed description of all the different species adapted for turning out, and of the various hybrids and crosses between them, is then given; and the work concludes with accounts of the ornamental species, such as the Gold, Silver, and Amherst pheasants, and the best methods of their management in aviaries.

Of the admirable engravings which illustrate the volume I may remark, in the words of Izaak Walton, "Next let me add this, that he that likes not the book should like the excellent pictures . . . . which I may take a liberty to commend, because they concern not myself."

W. B. TEGETMEIER.

NORTH FINCHLEY, N.

CONTENTS.

NATURAL HISTORY OF THE PHEASANTS.

CHAPTER I.
Habits, Food, Structure, &c. ... *page* 1

CHAPTER II.
Introduction, Distribution, &c. 21

MANAGEMENT IN PRESERVES.

CHAPTER III.
Formation of Coverts 41

CHAPTER IV.
Feeding in Coverts 51

CHAPTER V.
Rearing and Protection 58

MANAGEMENT IN CONFINEMENT.

CHAPTER VI.
Formation of Pens and Aviaries 77

CHAPTER VII.
Laying and Hatching ... ... *page* 95

CHAPTER VIII.
Rearing the Young Birds 109

DISEASES OF PHEASANTS.

CHAPTER IX.
The Gapes, Cramp, &c. 125

PHEASANTS ADAPTED TO THE COVERT.

CHAPTER X.
The Common Pheasant 143

CHAPTER XI.
The Prince of Wales's Pheasant ... 152

CHAPTER XII.
The Chinese Pheasant ... 155

CHAPTER XIII.
The Japanese Pheasant ... 161

CHAPTER XIV.
Sœmmerring's Pheasant 169

CHAPTER XV.
Reeves's Pheasant 177

PHEASANTS ADAPTED TO THE AVIARY.

CHAPTER XVI.
The Golden Pheasant 188

Chapter XVII.
The Amherst Pheasant ... ... *page* 199

Chapter XVIII.
The Silver Pheasant 206

Chapter XIX.
The Eared Pheasant 212

Chapter XX.
The Impeyan Pheasant 215

Chapter XXI.
The Argus Pheasant 220

APPENDIX.
Transport of Pheasants 227

LIST OF PLATES.

Prince of Wales's Pheasant (*P. principalis*)......	*Frontispiece*	
Common Pheasant (*P. colchicus*)	*To face*	143
Bohemian Pheasant (*P. colchicus—variety*) ⎫ Hybrid Pheasant (Reeves's and Bohemian) ⎭	,,	149
Chinese Pheasant (*P. torquatus*)	,,	155
Japanese Pheasant (*P. versicolor*)	,,	161
Sœmmerring's Pheasant (*P. sœmmerringii*)	,,	169
Reeves's Pheasant (*P. reevesii*)	,,	177
Reeves's Pheasant in Flight...........................	,,	185
Reeves's Pheasant (*P. reevesii*)	,,	183
Golden Pheasant (*Thaumalea picta*)	,,	189
Amherst Pheasant (*Thaumalea amherstiæ*)	,,	199
Silver Pheasant (*Euplocamus nycthemerus*)	,,	207
Eared Pheasant (*Crossoptilon mantchuricum*)......	,,	213
Impeyan Pheasant (*Lophophorus impeyanus*)	,,	214
Argus Pheasant (*Argus giganteus*)	,,	221
Argus Pheasant Displaying its plumage	,,	225

PHEASANTS
FOR COVERTS AND AVIARIES.

CHAPTER I.

NATURAL HISTORY OF THE PHEASANTS.

HABITS, FOOD, STRUCTURE, ETC.

THE PHEASANTS, properly so called (as distinguished from the allied but perfectly distinct groups which include the Gold and Silver pheasants, the Kaleege, the Monaul, &c.) constitute the genus or group known to naturalists under the title *Phasianus*. Of the true pheasants no less than thirteen distinct species have been described by Mr. D. G. Elliott, in his magnificent monograph on the *Phasianidæ*. Of these several are known only by rare specimens of their skins brought from scarcely explored Asiatic countries, and others cannot be regarded as anything more than mere local or geographical varieties of well known species. Since the publication of Elliott's *Phasianidæ* several additional species have been described.

Without including, however, such birds as have, from their rarity or other causes, no practical interest to English game preservers, there remain several well known species that will require our careful consideration. Such are: The common pheasant (*Phasianus colchicus*), now generally diffused

B

throughout southern and central Europe; the Chinese (*P. torquatus*); the Japanese (*P. versicolor*); the Reeves (*P. reevesii*); and the Sœmmerring (*P. sœmmerringii*). These, however, are so closely related in their structure, form, and habits, that their natural history and general management may be given once for all, and their distinctive peculiarities pointed out subsequently.

The pheasants constituting the genus *Phasianus.* are readily distinguished by their extremely elongated tail feathers, which attain their maximum development in the Reeves pheasant, reaching in that species to a length exceeding five or six feet. They are all destitute of feathered crests or fleshy combs, but are furnished with small tufts of feathers behind the eyes. In their native state they are essentially forest birds, frequenting the margins of woods, coming into the open tracts in search of food, and retreating into the thick underwood at the slightest cause for alarm. The common pheasant, which has been introduced from its native country, Asia Minor, for upwards of a thousand years, though spread over the greater part of Europe, and more recently introduced into America, Australia, and New Zealand, still retains its primitive habits.

"It is," says Naumann, in his work on the "Birds of Germany," "certainly a forest bird, but not in the truest sense of the term; for neither does it inhabit the densely wooded districts, nor the depths of the mixed forest, unless driven to do so. Small pieces of grove, where deep underbush and high grass grow between the trees, where thorn hedges, berry-growing bushes, and water overgrown with reeds, and here and there pastures and fields are found, are its chosen places of abode. Nor must well-cultivated and grain-growing fields be wanting where this bird is to do well. It neither likes the bleak mountain country nor dry sandy places; nor does it frequent the pine woods unless for protection against its enemies, or during bad weather, or at night."

"In our own country," says Macgillivray, "its favourite places of resort are thick plantations, or tangled woods by streams, where, among the long grass, brambles, and other shrubs, it passes the night, sleeping on the ground in summer and autumn, but commonly roosting in the trees in the winter."

Like the domestic fowl, which it closely resembles in its internal structure and its habits, the pheasant is an omnivorous feeder; grain, herbage, roots, berries, and other small fruits, insects, acorns, beech mast, are alike acceptable to it. Naumann gives the following detailed description of its dietary on the Continent. "Its food consists of grain, seeds, fruits, and berries, with green herbs, insects, and worms, varying with the time of year. Ants, and particularly their larvæ, are a favourite food, the latter forming the chief support of the young. It also eats many green weeds, the tender shoots of grass, cabbage, young clover, wild cress, pimpernel, young peas, &c., &c. Of berries: the wild mezereum (*Daphne mezereum*), wild strawberries (*Fragaria*), currants, elderberries from the species *Sambucus racemosa, S. nigra*, and *S. ebulus*; blackberries (*Rubus cæsius, R. idæus*, and *R. fruiticosus*); mistletoe (*Viscum album*); hawthorn (*Oratægus torminalis*). Plums, apples, and pears it eats readily, and cherries, mulberries, and grapes it also takes when it can get them. In the autumn, ripe seeds are its chief food, it eats those of many of the sedges and grasses, and of several species of *Polygonum*, as *P. dumetorum*; black bindweed (*P. convolvulus*); knot grass (*P. aviculare*); and also those of the cow-wheat (*Melampyrum*); and acorns, beech mast, &c., form a large portion of its food in the latter months of the year. Amongst forest plants, it likes the seeds of the hemp-nettle (*Galeopsis*), and it also feeds on almost all the seeds that the farmer sows."

To this long catalogue of its continental fare may be added the roots of the common silver weed (*Potentilla anserina*), and those of the pig-nut or earth-nut (*Bunium flexuosum,* and the tubers of the common buttercups (*Ranunculus bulbosus* and *R. ficaria*), which are often scratched out of the soil and eaten. Macgillivray states that "One of the most remarkable facts relative to this bird that has come under my observation, was the presence of a very large quantity of the fronds of the common polypody (*Polypodium vulgare*) in the crop of one which I opened in the winter of 1835. I am not aware that any species of fern has ever been found constituting part of the food of a ruminating quadruped or gallinaceous bird; and if it should be found by experiment that the pheasant thrives on such substances, advantage might be taken of the circumstance."

Thompson, in his "Natural History of Ireland," recounts the different varieties of food he observed in opening the crops of ten pheasants—from November to April inclusive. In seven he discovered the fruit of the hawthorn, with grain, small seeds, and peas. In one no less than thirty-seven acorns. Another had its crop nearly filled with grass; only one contained any insects, the period of examination being the colder months of the year; in summer the pheasant is decidedly insectivorous; all contained numerous fragments of stone. He also records that in the spring the yellow flowers of the pilewort (*Ranunculus ficaria*) are always eaten in large quantity, as are the tuberous roots of the common silver weed (*Potentilla anserina*), when they are turned up by cultivation. Mr. Thompson adds: "While spending the month of January, 1849, at the sporting quarters of Ardimersy Cottage, Island of Islay, where pheasants are abundant, and attain a very large size—the ring-necked variety, too, being common—I observed that these birds, in the outer or wilder coverts, feed, during mild as well as severe weather, almost wholly on hazel nuts. In the first bird that was remarked to contain them, they were reckoned, and found to be twenty-four in number, all of full size and perfect; in addition were many large insect larvæ. Either oats or Indian corn being thrown out every morning before the windows of the cottage for pheasants, I had an opportunity of observing their great preference of the former to the latter. I remarked a pheasant one day in Islay taking the sparrow's place, by picking at horsedung on the road for undigested oats."

Among the more singular articles of food that form part of the pheasant's very varied dietary may be mentioned the spangles of the oak so common in the autumn on the under side of the leaves. These are galls caused by the presence of the eggs of a species of gall-fly (*Neuropterus*) which may be reared from the spangles if they are collected in the autumn, and kept in a cool and rather moist atmosphere during the winter. About the fall of the leaf these spangles begin to lose their flat mushroom-like form and red hirsute appearance, and become by degrees raised or bossed towards the middle, in consequence of the growth of the enclosed grub, which now becomes visible when the spangle is cut open. The perfect insect makes its appearance in April and May. Some few years since Mr. R. Carr Ellison published the following account of their being eagerly sought after and devoured by pheasants in a wild state: " Just before the fall of the oak-leaf these spangles (or the greater part of them) become detached from it, and are scattered upon the ground under the trees in great profusion. Our pheasants delight in picking them up, especially from the surface of walks and roads, where they are most easily found. But, as they are quite visible even to human eyes, among the wet but undecayed leaves beneath the oaks, wherever pheasants have been turning them up, a store of winter food is evidently provided by these minute and dormant insects with their vegetable incasement, in addition to the earthworms, slugs, &c., which induce the pheasants to forage so industriously, by scratching up the layers of damp leaves in incipient decay which cover the woodland soil in winter. Not only have we found the spangles plentifully in the crops of pheasants that have been shot, but, on presenting leaves covered with them to the common and to the gold pheasants in confinement, we observed the birds to pick them up without a moment's hesitation, and to look eagerly for more."

The value of pheasants to the agriculturist is scarcely sufficiently appreciated; the birds destroy enormous numbers of injurious insects—upwards of twelve hundred wireworms have been taken out of the crop of a pheasant; if this number was consumed at a single meal, the total destroyed must be almost incredible. There is no doubt that insects are preferred to grain, one pheasant shot at the close of the shooting season had in its crop 726 wireworms, one acorn, one snail, nine berries, and three grains of wheat. Mr. F. Bond states that he took out of the crop of a pheasant 440 grubs of the crane fly or daddy longlegs—these larvæ are exceedingly destructive to the roots of the grass on lawns and pastures. As another instance of their insectivorous character may be mentioned the complaint of Waterton, that they had extirpated the grasshoppers from Walton Park. They also occasionally eat molluscous animals. Mr. John Bishop, of Llandovery, records that he killed a pheasant on the coast of Islay whose crop was filled with the coloured snails abounding on the bents or grass stems on the coast.

Lord Lilford, in his magnificent volumes on the "Birds of Northamptonshire," writes: "The pheasant, where not preserved in unreasonable numbers, is a good friend to the farmer, from the enormous number of wireworms and other noxious insects which it devours, to say nothing of its liking for the roots of various weeds; but it would be absurd to deny that grain forms its favourite food, and a field of standing beans will, as is well known, draw pheasants for miles. It is very much the fashion to feed the birds with maize; but, in our own opinion, the flesh of pheasants which have been principally fed upon this corn is very far inferior in flavour to that of those who have found their own living upon what the land may offer them."

Like their allies, the domestic fowls, pheasants are occasionally carnivorous in their appetite. A correspondent writes: "This morning my keeper brought me a pied cock pheasant, found dead (but still warm) in some standing barley. The bird was in finest condition, and showed no marks whatever, when plucked, of a violent death. On searching the gullet I extracted a short-tailed field mouse, which had doubtless caused death by strangulation." And a similar instance was recorded by Mr. Hutton, of Northallerton. The Hon. and Rev. C. Bathurst, in a letter published in *Loudon's Magazine of Natural History*, vol. vii., p. 153, relates that Sir John Ogilvy saw a pheasant flying off with a common slow-worm (*Anguis fragilis*); that this reptile does sometimes form part of the food of the pheasant is confirmed by Mr. J. E. Harting, who recounts, in his work on "The Birds of Middlesex," that "on examining the crop of a pied pheasant, shot in October, 1864, I was surprised to find in it a common slow-worm (*Anguis fragilis*) which measured eight inches in length. It was not quite perfect, having lost the tip of the tail; otherwise, if whole, it would probably have measured nine inches."

In October, 1888, Mr. J. B. Footner, of Tunbridge Wells, forwarded me a bottle containing three young vipers that were found with five others of equal size in the crop of a three parts grown hen pheasant, which he himself shot as a wild bird. Their length was slightly in excess of 7in., and the weight of the largest was exactly $\frac{1}{4}$oz. They were apparently young of the same brood. In his letter Mr. Footner recalled the fact that Sir Kenelm Digby, who lived in the time of Charles I., and married a lady of great beauty, used to feed his wife on capons fatted on young adders, which were believed to preserve beauty. Sir Kenelm Digby, whose portrait may be seen in Vandyke's Iconography, was remarkable as a charlatan, who proposed to cure wounds by applying a sympathetic powder to the weapons they were caused by, and who published a treatise on "Secrets pour la Beauté des Dames," from which the viper treatment is extracted.

The structure of the digestive organs of the pheasant is perfectly adapted to the assimilation of the food on which it feeds. The sharp edge of the upper mandible of the bill is admirably fitted for cutting off portions of the vegetables on which it partly subsists, and the whole organ is equally well adapted for securing the various articles of its extensive dietary. The food, when swallowed, passes into a very capacious membranous crop, situated under the skin at the fore part of the breast. From this organ portions gradually pass into the true digestive stomach, or proventiculus; this is a short tube, an inch and a half long, connecting the crop with the gizzard. Small as this organ may be, it is one of extreme importance, as the numerous small glands of which it mainly consists secrete the acid digestive or gastric fluid necessary to the digestion of the food; and in cases in which pheasants or fowls are fed on too great an abundance of animal food, or any highly-stimulating diet, this organ becomes inflamed, and death is frequently the result. From the proventiculus the food passes into the gizzard, which is lined with a dense thick skin; in its cavity the food is ground down to a pulp, the process being assisted by the presence of the numerous small stones and angular pieces of gravel, &c., swallowed by the bird. The food, thus ground to a pulp, passes on into the intestines, which are no less than six feet in length; in the upper part of this long canal it is mingled with the bile formed in the liver, the pancreatic fluid, &c., and, as it passes from one extremity to the other, the nourishment for the support of the animal is extracted; this being greatly aided by the operation of the two cæca, or blind intestines, which are very large in all the birds of this group.

The flight of the pheasant is strong, and is performed by rapid and frequent beats of the wing, the tail at the same time being expanded. The force with which the bird flies may be inferred from the result which has not unfrequently occurred when it has come into contact with thick plate-glass in windows. A correspondent states : "A few days ago, a cock pheasant rose about three hundred yards from my house and flew against the centre of a plate glass window, smashing it into a thousand fragments. The glass was 3ft. 8in. by 3ft. 4in., and ¼in. thick; and such was the force of the concussion that not a single piece remained six inches square.. A slight snow on the ground rendered the window more than usually a mirror reflecting the outer landscape. It is needless to say the bird was killed instantaneously. Two hen pheasants had on previous occasions been killed in the same way, but the glass was not damaged." Mr. G. A. Hackett, of Pailton House, Rugby, also wrote as follows: "I was much astonished to-day, at about two o'clock, by hearing a loud crash of glass in my smoking-room, and on going there I found a cock pheasant dead on the floor close to the window, and the plate of glass, which is 4ft. by 3ft. 6in., and ¼in. thick, in thousands of fragments. I am certain no blow from a man could have in like manner demolished the glass. The pheasant was a ring-necked, last year's bird, and weighed nearly 3lb." These instances occured in the day-time. Sometimes the birds are attracted by a light, as in the following cases: "On a very rough night in January, a hen pheasant flew through the hall window at Merthyr Manor, Bridgend, attracted by a light inside." And the following incident is related as occurring in a village not far from Bangor, on the banks of a river on the opposite side of which is a plantation well stocked with pheasants : "One stormy night there sat in a room of a small public, which had a window facing the plantation, six or seven men enjoying their pipes and beer, when all of a sudden crash went the window, out went the candle, and out rushed the men in great consternation. On examining the room a splendid cock pheasant was found under the table."

The wings, considered with reference to the size and weight of the bird, are short and small; from the secondary quills being nearly as long as the primary, they are very rounded in form, the third and fourth primary feathers being the longest. The wings are not adapted to a very prolonged flight, although the denizens of the wilder districts in the country fly with a speed and cover distances that are unknown to the over-fattened birds in our preserves. Long flights are, however, not altogether beyond the powers of the bird. One of unusual length was recorded by Mr. J. Cordeaux, of Great Cotes, Ulceby, who states that "when shooting in the marshes on the Lincolnshire side of the Humber, near Grimsby, a man who works on the sea embankment came to say that two pheasants had just flown over from the Yorkshire side, alighting within a few feet of where he was working among the rough grass on the bank. On going to the spot indicated, I at once found and shot them; they were both hens, and in very good condition. The Humber at this place from shore to shore is nearly four miles across. There was a strong northerly breeze blowing at the time, so that they would cross before the wind, or with the wind a little aslant. I have occasionally found pheasants in the marshes, and near the embankment, which I was sure must have come across, but had no direct evidence of the fact."

The comparatively small size of the wings necessitates their being moved with great force and velocity, and consequently the moving powers or muscles of the breast are very large and well developed, taking their origin from the deep keel on the breast bone. The tail is long, and tapers to a point; it is composed of eighteen straight pointed feathers.

The pheasant, like most of its congeners, is a terrestrial bird, seeking its food, making its nest, and rearing its young upon the surface of the ground. Its legs, like those of all true rasorial or scratching birds, are strong and muscular, consequently it is capable of running with great speed. The strong blunt claws are admirably adapted for scratching seeds and tuberous roots from the ground, or worms and larvæ from beneath fallen leaves.

Though seldom taking voluntarily to the water, the

pheasant is quite capable of swimming, as is proved by the following instances. A well-known game preserver writes: "When out walking to-day with my keeper, near the end of a long pond running under one of my woods, we fancied that we heard some young pheasants calling in the high grass. On going up to the place where we had heard the noise, an old hen pheasant got up and flew over the pond, which is about eighteen or nineteen feet wide at this place and about four feet deep. To our astonishment one of the young birds ran down to the water, went into it, and swam safely to the other side after its mother. The young birds could not have been more than fourteen days old." Old birds will also voluntarily swim across rivers, as in the following instance: "While flogging the waters of the Usk, I saw a sight that struck me with astonishment. A fine cock pheasant was walking about on the bank of the river, here quite thirty yards broad and running at the rate of four knots an hour. On our approach he quietly took to the water like a duck, and, after floating down stream a few yards, boldly struck across, and, swimming high and with great ease, reached the bank nearly opposite to the spot whence he set out." And other similar cases are on record, thus—Mr. Donald Campbell, of Dunstafforage, Oban, states: "Six pheasants, five cocks and a hen, attempted to fly across Loch Etive from one of the Ardchattan coverts on the north side of the loch, which near that spot varies from half a mile to a mile in width. When about half-way across one of them was seen either to fall or alight on the water, and its example was immediately followed by the other five. Fortunately, the son of the Ardchattan gamekeeper, who was in a boat on the loch at the time, observed the occurrence, and rowed to the spot; but as he had some distance to go, by the time he reached the birds they were very much exhausted and half drowned, and were drifting helplessly with the tide. He got them into the boat and took them ashore, and, after being well dried and placed in warm boxes near a good fire, they all eventually recovered.

The day was cold and frosty, and there was a slight fog on the water." When wounded and dropped into the water, pheasants swim with facility, and some instances are on record of their diving beneath the surface and rising at some distance.

As the breeding season approaches, the crow of the male, resembling the imperfect attempts of a young fowl, may be heard distinctly. It is followed, and not preceded as in the game cock, by the clapping of the wings; the pheasant and the domestic cock invariably reversing the order of the succession of these two actions. Like the domestic fowl, pheasants will also answer any loud noise, occurring either by day or night; they have been noticed replying regularly to the signal gun at Shorncliffe, which is fired at sunrise and sunset, and this in coverts situated some miles distant; and the practice with the heavy guns at the various military stations will often cause a chorus of "cucketing" in all the coverts for a great distance round.

The display of the plumage during courtship by the males varies in almost every species of gallinaceous birds. That of the pheasant was carefully described by the late Mr. T. W. Wood, in an interesting article on the "Courtship of Birds." Pheasants seem to possess no other mode of display than the lateral or one-sided method. In this the males disport themselves so as to exhibit to the females a greater number of their beautiful feathers than could otherwise be seen at one view. The peculiar attitude assumed by the male of the common species is correctly shown in the vignette on page 20 at the end of this chapter; the wing of the side nearest the female is partly opened and depressed, precisely in the same manner as performed by the male of the common fowl, and, in addition, the tail is expanded, and the upper surface turned towards the same side, whilst the bright vermilion skin around the eye is greatly extended, and the little purple aigrettes erected. Singular modifications of this sexual display of the plumage occur in the Argus and Golden

Pheasant and other species, which will be noticed in the chapters relating to those birds.

In a state of nature there is little doubt that the pheasant is polygamous. The males are armed with spurs, with which they fight, the stronger driving away the weaker, and the most vigorous propagate their kind.

The nest of the female is usually a simple hollow scraped in the ground. After depositing her eggs (usually about eight or nine in number) she is deserted by the male, and the task of incubation and rearing the young depends on her alone. The eggs vary in colour from a greenish brown to a greyish green; in size they are, on the average, an inch and five-sixths in length, by an inch and five-twelfths in width. The period of incubation is twenty-four days.

Hen pheasants, like common fowls, not unfrequently have nests in common, in which case as many as eighteen or twenty eggs will be found together. Sometimes three hens will take to the same nest, and as many as thirty eggs have been seen resulting from their copartnership. It is still more singular that the pheasant and the partridge often share the same nest. Mr Walter Yate, of Pemberton, Shropshire, stated, " About a week ago one of my workmen informed me that he had found a nest containing both partridge's and pheasant's eggs. I accompanied him to the place, and there saw the pheasant and partridge seated side by side with the utmost amity. I then had the birds driven off, and saw fifteen partridge's and sixteen pheasant's eggs laid indiscriminately together. The eggs were placed as though the nest had been common to both." Another correspondent writes : " About three weeks ago, when walking round a small wood belonging to me, and in which I usually breed a good sprinkle of pheasants, I discovered a partridge sitting on the edge of the bank of the wood; and when she went off to feed I was much astonished to find that she was sitting on nine pheasant's eggs and thirteen of her own, and, after sitting the usual time, hatched them all out." Mr. R. Bagnall-Wild records that "in June his keeper noticed three partridge nests, with thirteen, eleven, and eleven partridges' eggs, and four, two, and two pheasants' eggs, respectively in them. He carefully watched, and in all three cases found that the pheasants were hatched with the young partridges; and in September the young pheasants still kept with their respective coveys of partridges." Sometimes the hen pheasant, and not the partridge, is the foster parent. In the neighbourhood of Chesham, on the 6th of May, 1873, three pheasants' nests were observed to contain the following eggs :—the first, on which the hen was sitting, twenty-two pheasant's and two French partridge's eggs; the second, eleven pheasant's and five French partridge's eggs; and the third, six pheasant's and seven French partridge's eggs. Mr. W. D. Collins, of Cuckfield, records the fact that he found a grey partridge sitting on twelve of her own eggs, nine eggs of the red-legged partridge, and nine pheasant's eggs, all the three species having layed in the same nest. Mr. Higgins, of Hambledon, states that "A pheasant hatched out, in a piece of vetches of mine, seven partridges and five pheasants on July 6th. She sat on nine of her own eggs and eight partridge eggs." In some cases the nest is even of a more composite character, and the eggs of the common fowl, and those of partridges and pheasants, have all been found together; and instances have been recorded of wild hen pheasants laying in the nests of tame and also of wild ducks, and in the nest of the corncrake.

Although there is usually some attempt at concealment under covert, pheasants' nests are not unfrequently placed, even by perfectly wild birds, in very exposed situations. Mr. John Walton, of Sholton Hall, Durham, related the following account of the singular tameness of a wild bred bird: "A hen pheasant—a perfectly wild one so far as rearing is concerned, for we have no artificial processes here—selected as the site for her nest a hedge by a private cart road, where she was exposed to the constant traffic of carts, farm servants, and others, passing and repassing her quarters, all of which she took with infinite composure. She was very soon discovered on her nest, and actually suffered herself when sitting to be stroked down her plumage by the children and others who visited her, and this without budging an inch. In fact, she seemed rather to like it. Perhaps she became a pet with the neighbours from this unusual docility, and her brood (fourteen in number) was thereby saved; for every egg was hatched, and the young birds have all got safely away."

Habitually a nester on the ground, the hen pheasant will sometimes select the deserted nest of an owl or squirrel as a place for the disposition and incubation of her eggs. Several examples of this occurrence are on record, but the following may suffice to prove that the circumstance is not so unfrequent as may have been supposed. One correspondent writes as follows: "Our head keeper told me that one of his watchers had found a pheasant's nest up a spruce fir tree. I was incredulous, so I went with him, and had the under-man there to show us. The bird was sitting on the nest—an old squirrel's. The man said she had twelve eggs. He also told us that he knew of another in a similar situation in the same plantation. The nest 1 saw was about twelve feet from the ground. The watchers found it in looking for nests of flying vermin, as some had escaped the traps."

Another states: "A keeper on the Culhorn estate, when on his rounds in search of vermin, observed a nest, which he took to be that of a hawk, on a Scotch fir tree, about fifteen feet from the ground. On throwing up a stone out flew a fine hen pheasant. The keeper then ascended the tree, and found, to his astonishment, eight pheasants' eggs in an old owl's nest. He removed the eggs, and placed them under a hen, and at the expiration of three days he had eight fine lively pheasant birds."

A third states that "at Chaddlewood, near Plympton, Devon, a pheasant has built its nest (twelve feet from the ground) in a fork of an ash tree close to the house, and has now laid eight eggs."

It is difficult to ascertain whether or not in the instances in which the young are hatched in these elevated situations, they fall out of the nest and survive or are killed and carried away by predatory animals, or whether they are safely removed by the parent birds, and if so, by what means; even the following accounts do not throw much light upon the subject. A correspondent of *The Field* stated that "A hen pheasant made her nest in an oak tree, about nine feet from the ground. The young were hatched, and she succeeded in taking seven young ones safely to the ground, leaving five dead in the nest, and one bad egg." A second stated that in the park at Fillingham, Lincoln, a pheasant deposited eight eggs in the nest of a woodpigeon in a fir tree upwards of sixteen feet from the ground; she hatched out seven of them, but was unfortunate, as four were killed; they were supposed to have fallen from the nest. And a third reported that on the estate of the Marquis of Hertford, at Sudborne Hall, Suffolk, a pheasant had taken possession of a nest deserted by a sparrow-hawk, in a spruce fir, twenty-five feet from the ground, and hatched eight young ones, seven of which she succeeded in bringing safely down, but in what manner was not stated.

Although as a rule the male pheasant takes no heed of the eggs laid by the female, or of the offspring when hatched, there are some well ascertained exceptions. Wild cock pheasants have been seen sitting in nests in the coverts by perfectly credible witnesses; and, although it has been suggested that the birds might have been hens that had assumed the male plumage, such an occurrence is even more unlikely than that a cock should sit, for those hens are always perfectly barren, and must have assumed the male plumage at the previous autumnal moult; in this condition they have never been known to manifest the slightest desire to incubate. Cocks have also been known to protect the young birds, as in the following instance, which occurred in Aberdeenshire: "I have for the last fortnight almost daily watched a cock pheasant leading about a brood of young ones, whose mother has evidently come to grief. A more attentive and careful nurse could not be than this cock. He boldly follows his young charge on the lawns and to other places where he never ventured before, finds them food, and stands sentry over them with untiring perseverance. They are thriving so well under his care and growing so fast, that they will soon be able to shift for themselves."

The same singular occurrence has also taken place in an aviary. Lord Willoughby de Broke some time since published the following letter: "I have an aviary in which there is a cock pheasant and four or five hens of the Chinese breed; at the beginning of the laying season the cock scraped a hole in the sand, in which the hens laid four eggs; he then collected a quantity of loose sticks, formed a perfect nest and began to sit; he sat most patiently, seldom leaving the nest till the eggs were chipped, when the keeper, afraid of his killing them, took them from him, and placed them under a hen pheasant who was sitting on bad eggs; they were hatched the next day, and the young birds are now doing well."

Pheasants usually commence to lay in this country in April or May, the date varying somewhat with the season and the latitude; but, in consequence of the artificial state in which they are kept in preserves, and the superabundance of food with which they are supplied, the production of eggs, as in domesticated fowls often takes place at most irregular periods. Many instances are recorded of perfect eggs being found in the oviducts of pheasants shot during the months of December and January. For example, Sir D. W. Legard, writing from Ganton, Yorkshire, on the 27th of December, 1864, said: "At the conclusion of a day's covert shooting last Tuesday, a hen pheasant, which had been killed, was discovered by a keeper to have a lump of some hard substance in her; he opened her in my presence, when, to my astonishment, he extracted an egg perfectly formed, shelled, and apparently ready to be laid; it was of the usual size, but the colour, instead of being olive, was a greyish-white."

A nest containing an egg has been noticed as early as the 12th of March, and many cases are recorded of strong nests of young during the first few days of May. Lord Warwick's keeper, J. Edwards, in May, 1868, wrote as follows: "Yesterday (the 6th inst.), whilst searching for pheasant eggs in Grayfield Wood, I came upon a nest of thirteen pheasant eggs, twelve just hatched and run, and one left cheeping in the shell. The bird must have begun to lay in the middle of March, as they sit twenty-five days, and they do not very often lay (only every other day, at least at the commencement)." Other cases earlier by three or four days than this instance have been recorded. The Rev. G. C. Green, of Modbury, Devon, writes: "On Sunday, April 18, 1875, as my curate was returning from taking the duty in a neighbouring church, a hen pheasant started from the roadside hedge close to the town, and fluttered before him. While watching her movements he saw eleven young pheasants, apparently newly hatched, fluttering in the hedge, and at the edge of a pond close by. They soon scrambled into some cover, and the mother bird flew off to rejoin them from another quarter. I understand, from inquiry, that this is not a solitary instance of such an early brood of pheasants in South Devon."

On the other hand, examples of nests deferred until very late in the year are not unknown. Mr. W. W. Blest, of Biddenden, near Staplehurst, writes: "Whilst partridge shooting on the 3rd of September, 1874, we disturbed a sitting pheasant, the nest containing twelve eggs. We often hear of the early nesting of game birds, but rarely so late in the season." In October, 1869, Mr. Walter R. Tyrell, of Plashwood, near Stowmarket, forwarded to me a young pheasant, with the following letter: "When pheasant shooting with some friends yesterday, the 15th inst., in this neighbourhood, one of the beaters picked up dead, in a path in the wood we were in, a very young chick pheasant; it could not have been hatched more than a week. My keeper tells me he has found them (but very rarely) as young in September. I forward the young chick to you, in order that you may inspect it." I carefully examined the young bird, which was not more than two or three days old. These late-hatched birds were in all probability the produce of a second laying during the season.

The artificial state in which these birds exist, as supplied with nutritive food and protected in our coverts and preserves, leads to other departures from their natural conditions. Thus variations of plumage and size are much more frequent and more marked than would occur in the case of birds in a perfectly wild state. In some instances the size is very greatly increased. Hen pheasants usually weigh from two pounds to two pounds and a quarter, whilst the usual weight of cock pheasants is from about three pounds to three pounds and a half. Mr. Yarrell, in his "History of British Birds," mentions two unusually large; he says "The lighter bird of the two just turned the scale against four and a half pounds; the other took the scale down at once. The weights were accurately ascertained, in the presence of several friends, to decide a wager of which I was myself the loser." One of five pounds and half an ounce was sent me by Mr. Carr, of the Strand; this was a last year's bird of the common species. And in 1859 one bird, of the enormous weight of five pounds and three-quarters, was sent by Mr. Akroyd, of Boddington Park, Nantwich, to Mr. Shaw, of Shrewsbury, for preservation. Mr. Akroyd stated that "the bird was picked up with broken leg and wing forty-eight hours after the covert was shot, so had probably lost weight to some extent." In reply to the suggestion that it might possibly have been a large hybrid between the pheasant and the domestic fowl, Mr. Akroyd further stated "that the bird looked all its weight, and was as distinguished amongst its fellows as a turkey would be amongst fowls; yet it had no hybrid appearance whatever;" and Mr. Shaw stated that he weighed it several times. Moreover, he said, "the bird, had it been picked up when shot, would, I have little doubt, have weighed six pounds, there being nothing in its craw but two single grains of Indian corn; and when the length of time it remained wounded on the ground, with a broken thigh and wing, is taken into consideration, there can be little doubt of the fact." But the largest on record was described in vol. xlvi., p. 179, of *The Field*. G. C. G. writes: "I have received the following from Mr. Kelly in consequence of a discussion in *The Field* about the weight of a pheasant: 'Some few years since, while Admiral Sir Houston Stewart was residing at Ganton, he sent me a pheasant that weighed 6lb. wanting 1oz. He was an old bird, and the most splendid in form and plumage that I ever beheld. A few days afterwards being at Ganton, I told Sir Houston that I had weighed the bird, but I thought my weights must be incorrect, and asked him whether he knew its weight. He said, "You are quite right. I weighed it before I sent it to you, and that is my weight."'" In these cases of exceptionally large birds, it is usually found that the extreme weight is owing to the fattening influence of the maize on which they have been fed.

COCK PHEASANT DISPLAYING ITS PLUMAGE.

CHAPTER II.

NATURAL HISTORY OF THE PHEASANTS (CONTINUED).

NON-DOMESTICITY — INTRODUCTION INTO BRITAIN—DISTRIBUTION.

IT IS sometimes suggested by persons ignorant of the true nature of the pheasant, that it might be domesticated and reared like our ordinary farm-yard fowl. Such persons are apparently not aware that the instinct of domestication is one of the rarest possessed by animals. Man has been for some thousands of years capturing, subduing, and taming hundreds of different species of animals of all classes; but of these the number that he has succeeded in really domesticating does not amount to fifty. A very large proportion of animals are capable of being tamed, and rendered perfectly familiar with man; but this is a totally distinct state from one of domestication. The common pheasant is a good example of this distinction. Individual examples may be rendered so tame as to become even troublesome from their courage and familiarity; but although others have been bred in aviaries for many generations, their offspring still retain their original wildness, and when let out at large betake themselves to the woods and coverts as soon as able to shift for themselves. On the other hand, the allied species, the jungle fowl (*Gallus ferugineus*), the original of our domestic breeds of poultry, if reared in confinement, becomes immediately domesticated, the young returning home at night with a regularity that has given rise to the proverbial saying that " Curses, like chickens, come home to roost."

Examples of the tameness of individual pheasants are not rare; to the fearless nature of a sitting hen I have already alluded. The males become even more familiar, and even at times aggressive; one of the most amusing examples was recorded some time since by a correspondent, who wrote as follows: "Having recently been on a visit to a friend of mine living in Kent, I had an opportunity of there witnessing the effect of an extraordinary antipathy to crinoline exemplified in a fine cock pheasant which inhabited, or rather infested the grounds and shrubbery. He had been originally, I believe, reared on the premises, but had become as wild as any of his fellows, and, after having been lord of a harem of some seven or eight ladies last spring, who had all reared their families and gone off with them, had been left in loneliness, with his temper soured against the female sex at large. His beat was for about a quarter of a mile between the house and the entrance-gate, and on the approach of anything in the shape of crinoline his temper was roused to such a degree that he attacked it with all his might and main, flying up at the unnatural appendage, pecking fiercely with his bill, and striking out at it with his spurs like any game-cock. I witnessed all this with my own eyes, and was not surprised at the terror he had created among the females by whom he was positively dreaded, and not without reason. One lady had attempted to protect herself by taking a terrier as her guardian, who at first offered fight in her defence, but was soon compelled to show the white feather, and at the very sight of his antagonist ran off with his tail between his legs. At length, however, he met with his master in the shape of a gipsy-woman, who being of course uncrinolined, and therefore considering herself unjustly attacked, set upon him, and not only pulled out his tail, but crushed him with her foot, and left him on his back apparently in the agonies of death. The domestics, however, went to his assistance, and by their kind attentions he was restored. Still his old antipathy revived with his returning strength, and in a day or two the sight of crinoline again roused his wrath. Therefore, for fear of his meeting with an untimely end from some other strong-minded woman, it was decided that he should have his wing clipped, and be kept prisoner within the walls of the kitchen-garden."

The wife of Mr. Barnes (formerly head keeper to Mr. D. Wynham, of Denton Hall, near Salisbury) carefully nursed a very young hen pheasant with a broken leg. She got well, and in course of time was turned out with the rest of the brood into the adjacent woods. For several seasons afterwards this hen brought her own brood to the keeper's lodge.

Mr. T. B. Johnson, in his "Gamekeeper's Directory," mentions one he had reared from the nest that became uncommonly familiar : "It will follow me," he writes, "into the garden or homestead, where it will feed on insects and grass, and I occasionally observed it swallow large worms. Of all things, however, flies appear to be its favourite food. Before he was able to fly, I frequently lifted him into the window, and it was truly amusing to witness his dexterity in fly catching. He had been named Dick, to which he answers as well as possible. Dick is a very social being, who cannot endure being left alone; and if it so happen (as it occasionally does) that the bird finds every person has quitted the room, he immediately goes in search of some of the family; if the door be shut, and his egress thus denied, he utters the most plaintive noise, evidently testifying every symptom of uneasiness and fear in being separated from his friends and protectors. Dick is a great favourite, and on this account is suffered to take many liberties. When breakfast is brought in he jumps on the table, and very unceremoniously helps himself to bread, or to whatever he takes a fancy; but, different from the magpie or jackdaw under similar circumstances, Dick is easily checked. He is fond of stretching himself in the sunbeams : and if this be not attainable, before the kitchen fire. On being taken into the house he was presented to the view of the cat, the latter at the same time given to understand that the bird was privileged, and that she must not disturb him. The cat is evidently not fond of Dick as an inmate, but she abstains from violence. I have seen her, it is true, give him a blow with her paw, but this only occurs when the bird attempts to take bread, &c., from her; and not always then, as she frequently suffers herself to be robbed by him. Dick has also made friends with my pointers. He sleeps in my bedroom, but is by no means so early a riser as his fraternity in a state of nature; however, when he comes forth his antics are amusing enough; he shakes himself, jumps and flies about the room for several minutes, and then descends into the breakfast-room." Whether this bird would or would not have continued tame and domesticated during the following breeding season was unfortunately never ascertained, as it partook of the fate of most pets, and was killed accidentally by the opening of a door.

The incapacity of pheasants for domestication has been remarked by all those who have tried in vain to rear them as domestic birds. The late Mr. Charles Waterton, of Walton Hall, made the attempt under the most advantageous circumstances, and thus recounts the result of his experiments : " Notwithstanding the proximity of the pheasant to the nature of the barndoor fowl, still it has that within it which baffles every attempt on our part to render its domestication complete. What I allude to is, a most singular innate timidity, which never fails to show itself on the sudden and abrupt appearance of an object. I spent some months in trying to overcome this timorous propensity in the pheasant, but I failed completely in the attempt. The young birds, which had been hatched under a domestic hen, soon became very tame, and would even receive food from the hand when it was offered cautiously to them. They would fly up to the window, and would feed in company with the common poultry, but if anybody approached them unawares, off they went to the nearest covert with surprising velocity; they remained in it till all was quiet, and then returned with their usual confidence. Two of them lost their lives in the water by the unexpected appearance of a pointer, while the barndoor fowls seemed scarcely to notice the presence of the intruder; the rest took finally to the woods at the commencement of the breeding season. This particular kind of timidity, which does not appear in our domestic fowls, seems to me to oppose the only, though at the same time an unsurmountable, bar to our final triumph over the pheasant. After attentive observation, I can perceive nothing else in the habits of the bird to serve as a clue by which we may be enabled to trace the cause of failure in the many attempts which have been made to invite it to breed in our yards, and retire to rest with the barndoor fowl and turkey."

With regard to the date of the introduction of the pheasant into England, Mr. Thompson, writing in 1866, says he knows of no records which afford any clue to the period when it was first brought into this country; and that though probably its acclimatisation does not date back further than the Norman Conquest, yet it is possible that our Roman invaders may have imported it at a much earlier period, with other imperial luxuries.

Lord Lilford considers its introduction by the Romans as conclusively proved. In his "Notes on the Birds of Northamptonshire," he writes: "There appears to be no reason to doubt that the pheasant was introduced into England by the Romans, and the bird has now become so spread over most parts of Europe that it is almost impossible to say where it is really indigenous."

This suggestion is possibly near the truth, for the pheasant has been shown by Mr. W. Boyd Dawkins to have been naturalised in this country upwards of eight hundred years.

Writing to *The Ibis* for 1869 (page 358), that gentleman says: " It may interest your readers to know that the most ancient record of the occurrence of the pheasant in Great Britain is to be found in the tract ' De inventione Sanctæ Crucis nostræ in Monte Acuto et de ductione ejusdem apud Waltham,' edited from manuscripts in the British Museum by Professor Stubbs, and published in 1861. The bill of fare drawn up by Harold for the Canons' households of from six to seven persons, A.D. 1059, and preserved in a manuscript of the date of *circa* 1177, was as follows (p. 16):

> Erant autem tales pitantiæ unicuique canonico : a festo Sancti Michaelis usque ad caput jejunii [Ash Wednesday] aut xii merulæ, aut ii aganseæ [*Agace*, a magpie (?), *Ducange*], aut ii perdices, aut unus phasianus, reliquis temporibus aut ancæ [Geese, *Ducange*] aut gallinæ.

" Now the point of this passage is that it shows that *Phasianus colchicus* had become naturalised in England before the Norman invasion; and as the English and Danes were not the introducers of strange animals in any well authenticated case, it offers fair presumptive evidence that it was introduced by the Roman conquerors, who naturalised the fallow deer in Britain."

"The eating of magpies at Waltham, though singular, was not as remarkable as the eating of horse by the monks of St. Galle in the time of Charles the Great and the returning thanks to God for it :

> Sit feralis equi caro dulcis sub cruce Christi !

The bird was not so unclean as the horse—the emblem of paganism—was unholy."

But the conclusion that the pheasant was introduced into England before the Norman Conquest is not regarded as proved by those authorities who consider the tract " De inventione Crucis " as a miracle-mongering work that no cautious antiquary would accept as conclusive evidence.

In Dugdale's "Monasticon Anglicanum " is a reference

MEDIÆVAL HISTORY OF PHEASANTS.

by which it appears that the Abbot of Amesbury obtained a licence to kill hares and pheasants in the first years of the reign of King Henry the First, which commenced on the second of August, 1100; and Daniell, in his "Rural Sports," quotes "Echard's History of England" to the effect that in the year 1299 (the twenty-seventh of Edward I.) the price of a pheasant was fourpence, a couple of woodcocks three-halfpence, a mallard three-halfpence, and a plover one penny.

" To these notices," writes the Rev. James Davies in the *Saturday Review,* " might have been added another which seems to set the pheasant at a higher premium—to wit, that in 1170 Thomas à Becket, on the day of his martyrdom, dined on a pheasant, and enjoyed it, as it would seem from the remark of one of his monks, that 'he dined more heartily and cheerfully that day than usual.'"

Those who are interested in the subject will find a most interesting series of extracts respecting the mediæval history of this bird in Mr. Harting's "Ornithology of Shakespeare," from which we quote the following:

" Leland, in his account of the feast given at the inthronisation of George Nevell, Archbishop of York, in the reign of Edward IV., tells us that, amongst other good things, two hundred 'fesauntes' were provided for the guests.

"In the 'Privy Purse Expenses of Elizabeth of York,' under date 'the xiiijth day of Novembre,' the following entry occurs:

" 'Itm̃. The same day to Richard Mylner of Byndfeld for bringing a present of fesauntes cokkes to the Queen to Westminster... ... vs.'

"In the 'Household Book' of Henry Percy, fifth Earl of Northumberland, which was commenced in 1512, the pheasant is thus referred to:

" 'Item, FESAUNTES to be hade for my Lordes own Mees at Principall Feestes and to be at xijd. a pece.'

"'Item, FESSAUNTIS for my Lordes owne Meas to be hadde at Principalle Feistis ande to be at xijd. a pece.'*

"In the year 1536, Henry VIII. issued a proclamation in order to preserve the partridges, pheasants, and herons 'from his palace at Westminster to St. Giles-in-the-Fields, and from thence to Islington, Hampstead, Highgate, and Hornsey Park.' Any person, of whatever rank, who should presume to kill, or in any wise molest these birds, was to be thrown into prison, and visited by such other punishments as to the King should seem meet.

" Some interesting particulars in regard to pheasants are

---

* "As a copy of the 'Northumberland Household Book' is not readily accessible, we give the following interesting extract, showing the price, at that date, of various birds for the table :

Capons at iid. a pece leyn (lean).
Chickeyns at ½d. a pece.
Hennys at iid. a pece.
Swannys (no price stated).
Geysse iiiid. or iiiid. at the moste.
Pluvers id. or 1½d. at moste.
Cranys xvid. a pece.
Hearonsewys (i.e. Heronshaws or Herons) xiid. a pece.
Mallardes iid. a pece.
Woodcokes id. or 1½d. at the moste.
Teylles id. a pece.
Wypes (i.e. Lapwings) id. a pece.
Seegulles id. or 1½d. at the moste.
Styntes after vi. a id.
Quaylles iid. a pece at moste.
Snypes after iii. a id.

Perttryges at iid. a pece.
Redeshaukes i½d.
Bytters (i.e. Bittens) xiid.
Fesauntes xiid.
Reys (i.e. Ruffs and Reeves) iid. a pece.
Sholardes vid. a pece.
Kyrlewes xiid. a pece.
Pacokes xiid. a pece.
See-Pyes (no price).
Wegions at i½d. the pece.
Knottes id. a pece.
Dottrells id. a pece.
Bustardes (no price).
Ternes after iii. a id.
Great byrdes after iiii. a id.
Small byrdes after xii. for iid.
Larkys after xii. for iid."

This extract is especially interesting as throwing light incidentally on the condition of the country; the unreclaimed state of the land is shown by the abundance and cheapness of the wading birds. Woodcocks at a penny, and snipes at three a penny, contrast strongly with partridges at twopence and pheasants and peacocks at twelvepence each. Nor is the change in the degree of estimation in which the birds are now held less remarkable. Curlews, herons, and bitterns, which are now scarcely valued as edible, ranked equal to pheasants and peacocks, and were three or four times the value of a grouse, whilst a fishy sea-gull was worth two or three chicken or one woodcock.

furnished by the 'Privy Purse Expenses of King Henry VIII.'
For example, under date xvjth Nov. 1532, we have:

"'Itm̄ the same daye paied to the fesaunt
breder in rewarde ... ... ixs̄. iiijd.
"'Itm̄ the xxv daye paied to the preste the
fesaunt breder at Elthm in rewarde ij
corons ... ... ixs̄. iiijd.'

"And in December of the same year:

"'Itm̄ the xxijd. daye paied to the french
Preste the fesaunt breder for to bye
him a gowne and other necesarys ... xls̄.'

"From these entries it would appear that even at this date some trouble and expense was incurred in rearing pheasants. No allusion, however, is made to their being shot. They must have been taken in a net or snare, or killed with a hawk. The last-named mode is indicated from another source : *

"' Item, a Fesant kylled with the Goshawke.
"' A notice, two Fesants and two Partridges killed with the hawks.'

"As a rule they are only referred to as being 'brought in,' the bearer receiving a gratuity for his trouble.

"' Jany. 1536-7. Itm̄. geuen to Hunte
yeoman of the pultry, bringing to hir
gc̄e two qwicke (*i.e.* live) phesants ... vijs̄. vjd.
"' Ap^{l.} 1537. Itm̄. geuen to Grene the
ptrich taker bringing a cowple of
Phesaunts to my lady's grace ... ... iijs̄. ixd.
"'Jan. 1537-8. Itm̄. geuen to my lady
Carow's s'uñt bringing a quick
Phesaunt ... ... ijs̄.
"' Jan. 1543-4. Itm̄. geuen to Hawkyn,
s'uñte of Hertford bringing a phesant
and ptriches† ... ... iijs̄. iiijd.'

---

* "' Extracts from the Household and Privy Purse Accounts of the L'estranges of Hunstanton, 1519—1578.' (Trans. Roy. Soc. Antiq. 1833.)
† "' The Privy Purse Expenses of the Princess Mary, 1536—1544.' (Edited by Sir F. Madden, 1831.)"

"In a survey of the possessions of the Abbey of Glastonbury made in 1539, mention is made of a *'game' of sixteen pheasants* in the woods at Meare, a manor near Glastonbury belonging to the Abbey.

"The value set upon pheasants and partridges at various periods, as shown by the laws fixing penalties for their destruction, seems to have fluctuated considerably.

"By a statute passed in the eleventh year of the reign of Henry VIII. it was forbidden 'to take pheasants or partridges with engines in another's ground without licence in pain of ten pound, to be divided between the owner of the ground and the prosecutor.' By 23 Eliz. c. 10, 'None should kill or take pheasants or partridges by night in pain of 20s. a pheasant, and 10s. a partridge, or one month's imprisonment, and bound with sureties not to offend again in the like kind.' By 1 Jac. I. c. 27, 'No person shall kill or take any pheasant, partridge (&c.), or take or destroy the eggs of pheasants, partridges (&c.), in pain of 20s., or imprisonment for every fowl or egg, and to find sureties in £20 not to offend in the like kind.' Under the same statute, no person was permitted 'to buy or sell any pheasant or partridge, upon pain or forfeit of 20s. for every pheasant, and 10s for every partridge.' By 7 Jac. I c. 11, 'Every person having hawked at or destroyed any pheasant or partridge between the 1st of July and last of August, forfeited 40s. for every time so hawking, and 20s. for every pheasant or partridge so destroyed or taken.' Lords of manors and their servants might take pheasants and partridges in their own grounds or precincts in the daytime between Michaelmas and Christmas. But every person of a mean condition having killed or taken any pheasant or partridge, forfeited 20s. for each one so killed, and had to find surety in £20 not to offend so again."

For an early notice of the pheasant in Suffolk, namely in 1467, Mr. Harting has referred me to the household expenses of Sir John Howard, Knight, afterwards Duke of Norfolk, edited by Beriah Botfield for the Roxburgh Club, wherein (at p. 399) under date of April, 1467, at Ipswich, there is the entry: "Item xii. fesawntes pryse xii$_s$." He adds that there is apparently no earlier mention of the pheasants in Norfolk than some references in the accounts of the L'Estranges at Hunstanton in 1519, and the entry above quoted is the earliest for Suffolk.

In Essex, the pheasant is mentioned in a bill of fare, A.D. 1059 (as already noticed) and this is apparently the earliest allusion to the bird to be found in any part of England.

Mr. Harting further informs me that he has seen an ancient Psalter belonging to Lord Aldenham, in which there is a very fair coloured portrait of a cock pheasant, date A.D. 1260.

In Ireland, writes Mr. W. Thompson, in his natural history of that country, " The period of its introduction is unknown to me, but in the year 1589 it was remarked to be common." Fynes Moryson, who was in Ireland from 1599 to 1603, observes that there are "such plenty of pheasants as I have known sixty served up at one feast, and abound much more with rails, but partridges are somewhat scarce.''

In Scotland the pheasant does not appear to have been preserved at a very early period. Mr. R. Gray, in his work on "The Birds of the West of Scotland," says: " The first mention of the pheasant in old Scotch Acts is in one dated 8th June, 1594, in which year a keen sportsman occupied the Scottish throne." He might have been called "James the protector" of all kinds of game, as in the aforesaid year he " ordained that quhatsumever person or persones at ony time hereafter sall happen to slay deir, harts, pheasants, foulls, partricks, or other wyld foule quhatsumever, ather with gun, croce bow, dogges, halks, or girnes, or be uther ingine quhatsumever, or that beis found schutting with ony gun therein," &c., &c., shall pay the usual " hundreth punds," &c.

The distribution of the pheasant over Great Britain and

Ireland at the present time is very general, it being found in all parts of the kingdom where there is congenial shelter and some slight attempt at preservation and protection, without which it would soon be extirpated by poachers and its numerous natural enemies.

It is abundant even in the most populous counties, and is not at all uncommon in the immediate neighbourhood of the metropolis; but it is in the well-wooded and highly preserved districts of England that these birds most abound, and where they are excessively numerous. "The pheasant," writes Mr. Sterland, in his "Birds of Sherwood Forest," "abounds on all the estates in the forest district, and to such an extent that few would credit the immense numbers. They are almost as tame as barndoor fowls, and may be seen on the skirts of the various plantations. Carefully tended and fed, and all their natural enemies destroyed, they become so accustomed to the presence of man that in many parts they will hardly take the trouble to get out of the way, and are scarcely entitled to the appellation of wild. Under circumstances so favourable they multiply rapidly, but a natural limit seems to be set to their increase, and frequently, where they are most abundant, large numbers are found dead without apparent cause; these are always exceedingly fat and their plumage in the glossiest condition; they seem to drop down and die without a struggle. I have had them brought to me in this state, and have found their flesh plump and of good colour, and every feather smooth and perfect." I should rather incline to attribute the death in these cases to apoplexy, arising from over-feeding on maize and stimulating artificial food, than to any epidemic disease arising from overcrowding, as this attacks the young and destroys them long before they arrive at maturity.

"In Norfolk," writes Mr. Stevenson, in his admirable work on the birds of that country, "there are many portions where the pheasant exists in a perfectly wild state, and thrives well under the protection of the game laws, both soil and climate being alike favourable. It is in such districts, almost exclusively, that one still meets with the pure *Phasianus colchicus,* free from any trace of the ring-necked or Chinese cross in its plumage, but offering at the same time a poor contrast to those hybrid birds both in size and weight. Besides the thick undergrowth in woods and plantations, pheasants are particularly partial to low damp situations, such as alder and osier carrs, by the river side. In this country, also, stragglers from some neighbouring coverts are not unfrequently found on the snipe marshes surrounding the broads, where the sportsman, following up his dog at a 'running point,' is suddenly startled by the whirr of a noble 'long tail,' when never dreaming of any larger game than rails or water-hens."

In Scotland it is now very generally distributed in the western counties, from Wigtown in the south to Sutherland in the north. Mr. R. Gray writes: "In the neighbourhood of Loch Lomond, it may occasionally be noticed on the mountain sides, at a considerable elevation, sometimes as far up as twelve hundred feet. In Shemore Glen, I have seen male birds rise from the heath among the rocks, and, wheeling round, direct their flight down the valley with extraordinary speed. Very different indeed is the flight of these strong-winged natives of the glen from that of over-fed birds in wooded preserves; and as one bird after another shoots past in high air, one can hardly resist the impression that, if left to its own selection, the pheasant would adapt itself wonderfully to the drawbacks of its adopted country. Mr. Elwes informs me that he has frequently seen pheasants in Islay get up in the most unlikely places, such as an open moor, miles away from any covert or corn-field, and sometimes in a wet bog, where one would be more likely to find a snipe. On that island, where it was introduced about thirty years ago by Mr. Campbell, the pheasant is now not uncommon, and appears to be on the increase. In the Outer Hebrides it has likewise been introduced into Lewis by Sir James Matheson, who has obligingly informed me that, since its introduction twelve or fifteen years ago, it has become fairly established, although it has not increased to the extent that might have been expected in a more favourable locality. 'The deep drains in the peat moss,' writes Sir James, 'are supposed to be the cause of the death of the young chicks by their falling into them. For some years at first there was a want of covert for pheasants, but they are now better off in this respect, and are increasing gradually. Some of the first brood wandered about sixteen miles to the west side of the island, it is supposed in quest of covert.'"

The introduction of the pheasant into the northern districts of Scotland is, however, of comparatively recent date, for in the sixth edition of Moubray's "Domestic Poultry," 1830, it is stated : "In 1826, a solitary cock pheasant made his appearance as far north as a valley of the Grampians, being the first that had been seen in that northern region;" and my old friend, Andrew Halliday, told me that he remembered perfectly the introduction of the birds into the coverts near Banff belonging to the Earl of Fife, in which locality, Thomas Edwards, the Scottish naturalist, whose life has been so graphically written by Mr. Smiles, tells us it now seems to thrive very well, and is a beautiful ornament to parks and woods.

Messrs. Buckley and Harvie-Brown, in the "Fauna of the Orkney Islands," relate several unsuccessful attempts to introduce pheasants as wild birds into Orkney.

In Ireland it is also abundant, the common species being, according to Mr. Thompson, the well-known natural historian of the island, frequent in the various wooded parts, at least where it has been protected and preserved. "In the counties of Antrim and Down," remarks this writer, "the ring-necked variety—considered to have originally proceeded from a cross between the common and true ring-necked pheasant ($P.$ $torquatus$)—is not uncommon."

On the continent of Europe the pheasant is widely diffused throughout almost all the congenial localities in the south and central portions, where any effort is made in favour of its protection. In Scandinavia it has been successfully introduced; in 1867 we were informed by Mr. L. Lloyd, in his "Game Birds of Sweden and Norway," that it is not found, although attempts on a large scale were made to introduce it by the late King Oscar; but from the severity of the climate, and from the country swarming with vermin and birds of prey of all sorts, the experiment, in Mr. Lloyd's opinion, was not likely to be attended with success. Since that date the attempt has been successfully made by Baron Oscar Dickson, who, in 1873, reared seven or eight hundred birds. These have done well, for, in the *Morgenblad* of November 10, 1877, it is recorded that "Mr. (now Baron) Oscar Dickson and party shot in one day, on his property Bokedal, in Sweden, ninety pheasants, one deer, one hare, and one woodcock. There were five guns." And the same journal mentions that a brace of pheasants lived at full liberty on an estate in the neighbourhood of Christiania during the winter of 1876-7 without being fed or taken care of, and that they hatched in the summer of 1877, and reared four full-grown young ones. A brace more were let loose early in the spring of the same year, and also hatched and reared in the open. The first brace escaped from a pen, and nobody knew what had become of them. It was supposed that they were either frozen to death during the severe winter, had died of starvation, or had fallen an easy prey to foxes, cats, or hawks. But they survived, and found both shelter and food for themselves. Since that date they have increased rapidly, and on November 14 and 15, 1893, the Crown Prince shot over the Baron's preserves on the Island Wisingsö, in the Wetter Lakes, when 1548 pheasants were killed by six guns.

In New Zealand, the Great Britain of the southern hemisphere, the introduction of the pheasant has been a great success; so much so, that in a single season, that of 1871, six thousand birds were bagged in the immediate neighbourhood of the city of Auckland. Pheasants were first introduced into the province of Auckland about twenty years since, seven males and two females, the only survivors of two dozen shipped in China, comprising the original stock of the Chinese species. At the same time a number of the common species were liberated in another part of the colony. These were supplemented by six more Chinese birds in 1856. Both species have multiplied exceedingly, but their multiplication has in many places been lessened by the employment of phosphorised oats laid down to poison the rabbits.

The pheasant has also been introduced into several of the islands of the Pacific. By the kindness of Lieut. Ch. de Crespigny, of H.M.S. *Curaçoa*, I have received a specimen of the pheasants which are now breeding in the Samoan Islands. This pheasant is undoubtedly of the Chinese ring-necked species, the neck being nearly surrounded by the distinguishing white collar, but there is a considerable difference in the colour of the neck at the base and the scapular feathers, which are much lighter than in our ordinary species.

The Chinese pheasant was introduced by the Portuguese into the island of St. Helena in the year 1513, and has increased in numbers to a very considerable extent; but the present representatives of the original stock differ somewhat from their ancestors, both in the colour and markings of the plumage, as is described in the chapter on that species.

Very successful attempts have been made to introduce the different species of pheasants into North America as game birds, where in some parts they have become thoroughly acclimatised. The original stocks from whence the pheasants in the Western States were descended were imported direct from China, consequently the ring-necked pheasant (*P. torquatus*) is common in localities where the old English pheasant (*P. colchicus*) is almost unknown, although the latter has been introduced into the Eastern States on the Atlantic sea board.

In Oregon, where they were set at liberty in 1881, they have now become common, and they have spread and multiplied so well that complaints are made of their depredations in the grain fields.

The reports of the residents to the official inquiries are very interesting. Mr. Tyler, of Forest Grove, Oregon, writing in January, 1889, states :

"The females produce fifteen to eighteen eggs each litter, and hatch them all. . . . The old ones have lots of nerve, and will fight a hawk or anything that comes near them. The cocks will go into a barn yard and whip the best fowls we have, and run things according to their own notion. . . . Their favourite haunts are low grounds near the fields of grain, on which they depredate. . . . The golden pheasants have become numerous. Occasionally one is seen in our vicinity, about ninety miles from where they were turned loose four years ago; they are hardy, easily domesticated, but not so prolific as the ring-necks. Their flesh is white and tender."

A very good idea of the manner in which these species have succeeded in their new abode may be gathered from the circumstance that the farmers are shooting them as a nuisance, as they destroy the wheat. An interesting fact is that the gold pheasant (*Thaumalea picta*)—kept in England only as an ornamental aviary bird—has become wild in Oregon, and the Americans have found its flesh white and tender. I have eaten gold pheasants that had run wild in this country, and can fully indorse the statement. I have often wondered that some landed proprietor, living in a suitable locality bordering on woods and coverts, to whom beauty was of the first consideration, had not attempted to rear the gold pheasant in the open. The birds can be bred in a wild state, and yet remain so fearless as to come and feed from the hand; and it would be difficult to imagine any more gorgeous ornament to a country house than would be afforded by these birds.

Nevertheless, there is a much more beautiful bird than even the golden pheasant, and that is the cross between it and the Amherst pheasant (*T. amherstiæ*). This is not a sterile hybrid, but is perfectly fertile, either *inter se*, or with either of the parent races. For breeding in the open, it would be found hardier than either of the pure breeds from which it is descended, and, as it is larger than the golden pheasant, would make a better bird for the table, should anyone think of killing and eating an object of such surpassing beauty.

In the Eastern States the pheasants are in certain localities doing very well; as many as a thousand birds have been reared and turned out by a single keeper, and the pheasant is generally regarded as the future game bird of the country, as it can stand not only the severe heat of summer, but the cold and blizzards of the winter. A number of game clubs have been formed for their protection, and large numbers are raised in the Long Island preserves. They are also extending in several parts of New Jersey, New York, and Vermont. The Game Commissioners of Ohio are encouraging their breeding, and, to quote the words of the *Boston Herald*, "the outlook for the handsomest and most delicious game bird in the world is quite rosy in this country."

In the countries nearest to the locality from whence the common pheasant is supposed to have been derived, it is not, strange to say, abundant; thus the Rev. H. B. Tristram informs us that it does not appear to be known in Syria. In Greece, the Hon. T. L. Powys, writing in *The Ibis*, informs us that "The only localities in which I have seen pheasants in these parts were once on the Luro river, near Prevesa, in March, 1857, on which occasion I only saw one, the bird having never previously been met with in that part of the country; and again in December of the same year, in the forests near the mouth of the river Drin, in Albania, where it is comparatively common, and where several fell to our guns. In this latter locality, the pheasant's habitat seems to be confined to a radius of from twenty to thirty miles to the north, east, and south of the town of Alessio—a district for the most part densely wooded and well watered, with occasional tracts of cultivated ground, Indian corn being apparently the principal produce, and forming, with the berries of the privet (which abounds throughout Albania), the chief food of the present species. We heard many more pheasants than we saw, as the woods were thick and of great extent, our dogs wild, and we lost a great deal of time in making circuits to cross or avoid the numerous small but deep streams which intersect the country in every direction. This species is particularly abundant on the shores of the Gulf of Salonica, about the mouth of the river Vardar; and I have been informed, on good authority, that pheasants are also to be found in the woods of Vhrakori, in Ætolia, about midway between the gulfs of Lepanto and Arta." With regard to the present distribution of the species, Mr. Gould, in his "Birds of Asia," states that the late Mr. G. T. Vigne shot it in a wild state at the Lake of Apollonia, thirty-five miles from Broussa, to the south of the sea of Marmora, and that the late Mr. Atkinson found it on the Kezzil-a-Gatch and the country to the west of the river Ilia. Mr. C. G. Danford, in his notes on the ornithology of Asia Minor, writes: "The English Consul, Mr. Gilbertson, informed us that pheasants, though generally becoming scarce, were still common near Lake Apollonia, where a couple of guns had last year killed over sixty head in two or three days shooting." (*Ibis*, 1880, p. 98.)

Lord Lilford, writing in 1895, states: "The only country in which we have personally met with it in an unpreserved and perfectly wild state is on the shores of the Adriatic, near Alessio, in Albania, where it is, or was, by no means uncommon in the low-lying forest country near the mouth of the river Drin; it is also to be found in considerable numbers near Salonica and in certain other localities in European Turkey. But the best authorities seem to agree that the true home and headquarters of the species are the shores of the

Caspian, the valleys of the Caucasus, and Northern Asia Minor. Very closely allied forms, however, are to be met with from the Caspian, through Asia, to the shores and islands of China."

Professor H. H. Giglioli, writing of Corsica, states: "I was repeatedly assured of the presence in the island, among the hills of Aleria on the eastern coast, of the pheasant (*Phasianus colchicus*) in a perfectly wild condition. I see that Mr. Jesse reports the same thing. . . . I am still making inquiries on the subject; but, as far as I can see, no record of its introduction by man is forthcoming." (*Ibis*, April, 1881.)

The vignette represents the head of a hen pheasant with a singular beak, the upper mandible having passed between the sides of the lower jaw. The bird was found dead from starvation. It is evident that the deformity was caused by the bird flying against a branch, the blow bending the upper mandible and causing it to pass between the rami of the lower.

CHAPTER III.

MANAGEMENT OF PHEASANTS IN PRESERVES.

FORMATION OF COVERTS.

BEFORE any satisfactory progress can be made in the preservation of pheasants, the existence of good and well-protected coverts is indispensable; and where these do not naturally exist, the very first action of the game preserver must be to effect their plantation on a scale commensurate with his desires. This necessarily cannot be done without expense, but a large stock of pheasants cannot be secured, save under the most exceptional circumstances, without a very considerable outlay.

Some years since the subject of the formation of coverts for pheasants was discussed in a very exhaustive manner in the columns of *The Field*, and some admirable practical letters, detailing the experiences of the writers, appeared in that paper; these are worthy of the most attentive consideration, and I have great pleasure in availing myself of the opportunity of quoting from them. One of the most practical of the writers, the late Mr. R. Carr Ellison, of Dunston Hill, Durham, strongly advocated the formation of pheasant roosts of spruce and silver firs, as affording the birds absolute security against the attacks of night poachers. He writes:—" A number of country gentlemen who do not consider field sports of primary importance, feel it right to abstain from the preserving of pheasants. They see that the temptation which these birds offer, when perched upon naked larches and other trees, at night, is too strong to be resisted by many a lad or working man in the vicinity, who, but for this particular allurement to evil, might go on respectably and quietly enough. They know that their duty towards their own sons is to keep them out of needless temptations, and they are unwilling to expose the sons of other and poorer men to trials which experience shows they too often cannot resist. Some have forbidden all night watching of these birds, trusting them entirely to the protection of the pines and firs scattered in their plantations, in the branches of which it is impossible for any one to see the pheasants which happen to select them as a roosting-place. Now, I have for twenty-two years preserved these birds in very considerable numbers without any night watching, and in a country where all my neighbours have been repeatedly visited by gangs of poachers coming sometimes from considerable distances, as well as by occasional depredators of the vicinity. I resolved to reject all night watching, and one of the first things that I did, as a very young man, was to plant ten acres of spruce fir and Scotch pine in a central and sheltered part of the estate, which might serve as an impregnable roosting-place for pheasants. This was thirty years ago and more. At ten years of age, the plantation was already of great service, and at fifteen was invaluable. As it has been regularly thinned, it is now as good as ever. A number of birch-trees were intermixed, which were very useful in drawing up and hastening the growth of the spruces without exhausting the soil, as too great a multitude of firs would have done. Nor do the pheasants resort to the birch at night as they do to some other trees, larch especially, because they find that its branches are not sufficiently horizontal to afford commodious perches.

"Ten years later I formed a second pheasant-roost of two acres in extent, very near my house, and of this I have had the full benefit for many years past. It is generally full of pheasants, and not one of them is visible to the keenest eye in the clearest moonlight. It consists of spruce and silver fir, regularly and unsparingly thinned to keep the trees in health and vigour. We never think of night watching, even though guns be heard on adjoining estates, and the poachers have long given us up in despair. This lesser stronghold is kept sacred from the guns of sportsmen, who are sure to find the cock pheasants dispersed through all the other plantations during the daytime. The first thing the birds do on a winter's morning, after pecking up a few beans near their roost, is to wander in search of their natural wild food in the woodlands, of which food the tuberous root of the celandine, or wood-ranunculus, forms here a principal part. But, besides the remains of acorns and beech-nuts, they feed, I believe, much on the fallen keys of the ash and sycamore, on hips and haws, and on tender blades of grass, besides innumerable worms, eggs of slugs, and larvæ of insects. Tempted by these dainties, and in frosty weather even by the crisp green leaves of the holly, the cock pheasant will leave his beans and barley, and betake himself to freer haunts every fine day, and there the sportsman will find him; but, if his life be spared, he seldom fails to return at night to his warm roost among the spruces, only with the advance of spring will he quit it; for habit has made him luxurious as to his nights' quarters, and more sensitive of cold than less lucky pheasants.

"The Scotch pine is not nearly so tempting to the pheasant at night as the spruce and silver firs, because its branches are not sufficiently horizontal; yet, on dry hungry soils, it must be largely intermixed, since the firs are not to be depended on to flourish on such ground. In some cases, a stronghold may be formed entirely of hollies, Portugal laurels, and yews. For hen pheasants it will be excellent; but the cocks, which prefer to roost higher, should have a few firs or pines close at hand for their accommodation. All food should be given in or near to these secure nocturnal retreats."

Respecting the conversion of existing mixed plantations into night coverts for pheasants, the same gentleman remarks that "any plantation containing a due proportion of pines, or of spruce and silver fir, can be readily made a secure roosting-place for pheasants, if conveniently situated for the purpose, and not too much exposed to violent winds. All that is necessary is to cut out the larches as rapidly as can be done without letting in the wind too suddenly. The oaks, ashes, beeches, &c., may be allowed to stand wherever they do not injure a thriving pine or fir. The larches only are a dangerous temptation to the pheasants at roosting time. Their perfectly horizontal branches, and the considerable amount of shelter which their numerous twigs and regular head afford to the birds, induce many to perch in them; whereas young oaks, ashes, &c., attract very few indeed. If the plantation consisted entirely of resinous trees, so that none of the last-mentioned hardwood trees are present, then we have to consider what is to be done to fill up the vacancies. If the soil be tolerably moist and fertile, I would recommend that all the larger openings be filled with the best and strongest plants of silver fir that can be procured—say from two to three feet in height. Let a cluster of three or more of these be planted in pits, carefully prepared with spade and pickaxe, about five feet asunder, in the centre of every opening; for it is a pity to waste such plants in closer proximity to tall pines and spruces. If there be room for only one silver fir, let only one be planted. This species is not very liable to be nibbled by hares and rabbits if protected for the first year. Let the branches of the felled larches, with which the ground must still be half covered, be drawn around these young plants without delay, for very little will suffice to turn the enemy aside.

"Silver firs are very preferable to spruces or pines for filling up vacancies, for these latter, when drawn up slender by shade and shelter, are sure to be ruined by hares and rabbits, whereas the silver fir is of a different habit, and will not be drawn up in the same manner, nor is its taste so attractive to the marauders. It also bears being removed large from the nursery, with very little injury or check to its growth. Consequently, large plants of it, with earth adhering, though somewhat costly, are well worth their price to the planter who knows where and how to use them. Around these, and nearer to the tall pines and spruces, may be tried plants of the holly-leaved berberis and common laurel, which may not improbably succeed. Immediately under the pines and spruces it is useless to plant anything. The only covert to be obtained there is from heaps of branches left upon the ground as often as the trees are thinned. And this should be done almost annually, to ensure plenty of room to the best and most thriving amongst them, whose side branches will then gradually become more or less pendulous, and so will afford far more shelter than could be obtained from a larger number of trees standing too thick. Pheasants in a covert like this need no great quantity of shelter upon the ground, for they sit, even during the daytime, chiefly in the tree-tops. They bask there, on the south side of the summit of a spruce or pine, in the sun's rays, with great delight; and in heavy snow-storms whole days will often pass when they never descend to feed, but prefer to sit quiet, eating the green spines of these resinous trees (in the manner of the black grouse and capercailzie) when crispened by the frost, and depending upon snow by way of beverage. I have strongly advocated the spruce and silver firs as affording the most tempting perch to the birds at nightfall; still, be it understood, that the Scotch pine, pinaster, Weymouth pine (*P. laricio*) and others are all excellent. All that is needed is a little generalship and foresight in pheasant preservers, and a determination to confide in these resources, rather than in the expensive, dangerous, and inefficient practice of employing night watchers."

Commenting on these suggestions, another correspondent writes: "I am not aware that the practical advantages and excellence of the plan of planting large clumps or squares of spruce, either alone or blended with silver firs, and mixing, or not, a few deciduous trees with them, for the special purpose of forming pheasant roosts, have ever been so fully and perspicuously set forth as explained in the previous article. I could quote an instance of extensive coverts having been planted on a similar principle, save that oaks were planted in lieu of birch, with the ultimate view of affording these birds the opportunity of preening their plumes whilst perched on the topmost boughs, and enjoying themselves in this secluded retreat during bright weather, to which luxury, under such circumstances, they are very partial. In these cases the Spanish chesnut tree might sometimes perhaps be found an eligible substitute for either the birch or the oak. The larch undoubtedly is a favourite roosting tree with the pheasant, so much so indeed that I have seen odd ones roosting in larches growing within a few yards only of the impenetrable spruce grove. Besides being horizontal, the branches of the larch are rough, affording good foothold, and when the tree is properly grown are but at short distances one above the other, whilst, the collaterals being numerous, the tree in reality affords far more shelter than it appears capable of yielding, though, of course, far too little to conceal the bird from the prying eye of the night poacher. Pheasants are remarkably fond of 'hips;' and if the wild rose tree which produces them be kept low by a proper attention to pruning, not only can the birds reach the fruit easily, but the branches stool out and afford admirable covert. Cock pheasants are naturally of a vagrant turn, and at times will 'leave their beans and barley,' in order to indulge in this their favourite propensity to rove in search of their natural wild food in the woodlands, hedgerows, &c. Early in December last I received a brace of remarkably fine young cock pheasants shot on a manor where the best artificial food is abundantly provided, yet the crop of one of them contained ten full-sized acorns. Apart, too, from their utility as being by far the warmest, most sheltered, and the only thoroughly poacher-proof night coverts for these timid birds, which at roosting time usually court the densest sylvan shade—these evergreen groves possess the signal advantage of harmonising well with, and adding singular beauty to, the surrounding scenery; whilst the internal gloom—*lucus a non lucendo*—pervading them, has also its own peculiar charms, though it be of a sombre character."

It may be remarked that evergreen night coverts are not so essential south of the Trent, owing to the vigorous growth of underwood in the southern counties, which renders it almost impossible for poachers to traverse the coverts by night, even during bright moonlight; so that pheasants roosting on deciduous trees are much safer than they would be in the north, where underwood is comparatively feeble and scanty.

Writing to me on this subject, Mr. Carr Ellison added: "In the extreme north of England, and in Scotland, underwood of bramble grows feebly, except along warm southerly slopes. Nevertheless nature introduces another covert plant of great value, which fears neither cold shade, nor open and windy exposure—namely, the native tussock grass of mooredges and upland pastures, *Aira cæspitosa*, popularly called "*bull-fronts*," of which most of our exposed woodlands are full. It is easily transplanted, or propagated by seed, on which latter both pheasants and black game feed. It is a favourite covert for hares, affording perfect protection from the cold winds that sweep through plantations destitute of underwood, like too many in the north.

"Yet these apparently unpromising strips or clumps of bare stems are often frequented by fine broods of self-reared pheasants, thanks to the bull-fronts and bracken."

If it be desired to see the pheasants in the neighbourhood of the mansion, it should be borne in mind that the shrubberies of rhododendron so frequently seen skirting lawns and pleasure grounds are not frequented by pheasants like those of yew, holly, and privet, chiefly because no fallen berries are to be found underneath them. But if a handful of barley, peas, or beans be thrown from time to time among the more open and taller rhododendrons, the pheasants will soon learn to resort to them, after which some of the same fare may be cast into the thicker parts, where the birds will soon find it. In this way our beautiful rhododendron thickets near the garden and mansion may be utilised for pheasants more than heretofore.

The late Mr. Charles Waterton, who protected every bird in his domain, published the following details of his method of preserving the pheasants at Walton Hall :—" This bird has a capacious stomach, and requires much nutriment, while its timidity soon causes it to abandon those places which are disturbed. It is fond of acorns, beech mast, the berries of the hawthorn, the seeds of the wild rose, and the tubers of the Jerusalem artichoke. As long as these, and the corn dropped in the harvest, can be procured, the pheasant will do very well. In the spring it finds abundance of nourishment in the sprouting leaves of young clover; but from the commencement of the new year till the vernal period, their wild food affords a very scanty supply, and the bird will be exposed to all the evils of the Vagrant Act, unless you can contrive to keep it at home by an artificial supply of food. Boiled potatoes (which the pheasant prefers much to those in the raw state) and beans are, perhaps, the two most nourishing things that can be offered in the depth of winter. Beans in the end are cheaper than all the smaller kinds of grain, because the little birds, which usually swarm at the place where pheasants are fed, cannot swallow them; and, if you conceal the beans under yew or holly bushes, or under the lower branches of the spruce fir tree, they will be out of the way of the rooks and ringdoves. About two roods of the thousand-headed cabbage are a most valuable acquisition to the pheasant preserve. You sow a few ounces of seed in April, and transplant the young plants 2ft. asunder, in the month of June. By the time that the harvest is all in, these cabbages will afford a most excellent aliment to the pheasant, and are particularly serviceable when the ground is deeply covered with snow. I often think that pheasants are unintentionally destroyed by farmers during the autumnal seed-time. They have a custom of steeping the wheat in arsenic water. This must be injurious to birds which pick up the corn remaining on the surface of the mould. I sometimes find pheasants, at this period, dead in the plantations, and now and then take them up weak and languid, and quite unable to fly. I will mention here a little robbery by the pheasants, which has entirely deprived me of a gratification I used formerly to experience in an evening's saunter down the vale. They have completely exterminated the grasshoppers. For the last fourteen years I have not once heard the voice of this merry summer charmer in the party. In order to render useless all attempts of the nocturnal poacher to destroy the pheasants, it is absolutely necessary that a place of security should be formed. I know of no position more appropriate than a piece of level ground at the bottom of the hill, bordered by a gentle stream. About three acres of this, sowed with whins, and surrounded by a holly fence to keep the cattle out, would be the very thing. In the centre of it, for the space of one acre, there ought to be planted spruce fir trees, about 14ft. asunder. Next to the larch, this species of tree is generally preferred by the pheasants for their roosting-place; and it is quite impossible that the poachers can shoot them in these trees. Moreover, magpies and jays will always resort to them at nightfall; and they never fail to give the alarm on the first appearance of an enemy. Six or seven dozen of wooden pheasants, nailed on the branches of trees in the surrounding woods, cause unutterable vexation and loss of ammunition to these amateurs of nocturnal plunder. Small clumps of hollies and yew trees, with holly hedges round them, are of infinite service, when planted at intervals of one hundred and fifty yards. To these the pheasants fly on the sudden approach of danger during the day, and skulk there till the alarm is over." It is sometimes desirable to supply the want of ground covert for young birds in fir plantations where there is only short grass. The readiest mode of doing this is to use the trimmings of hedges, boughs, and tops of trees; the latter should be cut about a yard long and stuck in holes made with a crowbar. The high grass soon grows in amongst the sticks, and makes very good ground covert, which will last some years; or the roots of young spruce trees may be cut on one side, when the trees may be pulled down into a nearly horizontal position, and kept so by filling up the hole with the earth dug out.

The vignette represents the head of a pheasant in which the upper mandible had been shot away; nevertheless, the bird when killed was in good condition.

CHAPTER IV.

MANAGEMENT OF PHEASANTS IN PRESERVES (CONTINUED).

FEEDING IN COVERTS.

THE FOOD necessary to keep together a large stock of pheasants during the winter months, and prevent them straying to adjoining preserves, may be supplied in various modes. The birds may either be hand-fed day by day in the same manner as domestic fowls; or from troughs which are so constructed as to prevent the food being accessible to smaller birds; or they may be supplied with small stacks of unthrashed corn, from which to help themselves.

"If fed by hand, a fixed place is necessary, to which the pheasants must be accustomed to resort at a particular hour, otherwise the sparrows and other small birds will have far more than their fair share of the grain, particularly in severe weather when the ground is frozen hard. Fed in this manner, the birds become almost as tame as farm-yard fowls. In order to accustom them to one spot, at the end of September or earlier, according to the season, carry a few bundles of beans and barley, in the straw, to the spots in the coverts which are selected for feeding places; by watching these bundles it will be soon found when they have attracted the notice of the birds, and when it is observed that they have been attacking them, the better plan is to pull them apart, so as to enable the corn to be found more readily. When the corn is beginning to decrease, feed from the hand, daily; and, in order to ensure regularity, allow one man to distribute at the feeding-place, among the decaying barley-straw and beanhaulm, a small bagful of beans and barley, as early as he can find his way to the spot in the morning, concealing the corn as well as he is able; later in the day, say towards three or four in the afternoon, again deposit a mixture of barley and white peas, concealing the corn as before. In this way scarcely a grain of corn is lost. Woodpigeons and jays will sometimes intrude; but, with attention in concealing the corn, and punctuality in feeding, any waste worth notice may be prevented, and by observing how many birds come up to their food, it is easy to discover when anything is going wrong, as the least disturbance will make pheasants shy, and will be enough to put the keeper on the alert to discover the cause."

When fed by hand in this manner, a great variety of food may be used. Maize is certainly one of the best; weight for weight it is usually much cheaper than barley, is better relished by the pheasants, is far more fattening, and it possesses the great recommendation of not being so readily devoured by the sparrows, especially if the large coarse and cheaper varieties are purchased. A correspondent, who has kept pheasants for many years, and taken much trouble to ascertain their preference for different kinds of food, states, as the result of his experience, that "they prefer maize or Indian corn to any other food that can be given to them. I have frequently given the pheasants that come regularly to my window to be fed equal parts of Indian corn, peas, small horse-beans, wheat, barley, and oats, and they invariably take them in the order in which I have written them. I have also frequently done the same with those I keep shut up for laying, and always with the same results. Pheasants that I have had from elsewhere to put with them in confinement, and that have never seen maize, take to it in a couple of days, and then, like the others, will eat nothing else so long as they can get it; and if I try them with the mixture above named I find all the other grain neglected. The young pheasants at the coops begin to eat it before they are as large as partridges, and then entirely neglect the barley, &c. I never see pheasants that are kept up in better condition than my own, and they have nothing but Indian corn, a few turnip leaves, and clods of turf to pull to pieces. Another great advantage of maize is that small birds cannot steal it, with the exception of the tomtit, and though almost the smallest he holds the corn with one foot and hammers away like a miniature woodpecker, commencing at the part of the grain that is attached to the stalk, finding that the only road in. It is but a very small part of each corn that he is able to eat, but it seems to possess great attraction for him. There are six or eight of these little birds live constantly near my house at this season; and though chaffinches, blackbirds, and thrushes all try their best at the maize, they soon give it up hopelessly. Rooks take it greedily, and were it not for an occasional ball from the air gun they would rob the pheasants of every grain."

In feeding pheasants in this manner, care should be taken to change the ground frequently, for if they are fed on the same place for a continuance the ground becomes tainted, the food is necessarily soiled by the excrements of the birds, and disease is the invariable result.

Feeding troughs, which open with the weight of the pheasant when standing on an attached bar in front of the corn, are not extensively used. The objections to them are, in the first place, their expense, some fifteen shillings to thirty shillings each, which becomes a serious item when many are required; their liability to get out of order; and, lastly, the unlimited supply they afford to the feeding bird, which crams itself to repletion without any exercise, and is disinclined to seek food on its own account.

Unquestionably, the best mode of feeding pheasants is by

the use of small stacks of unthreshed grain or beans; but even this may be done in a wrong as well as a right manner. The late Mr. W. Lort, an enthusiastic practical sportsman, made the following suggestions: "Pheasants may be easily fed from small thatched stacks made with bundles of different kinds of grain. The only operation then required—pulling a bundle or two from the stack and cutting the bands—may be performed every two or three days; though, by the way, I must say I like someone to see my pheasants every day; and those who want game will find it to their interest to have it well attended to. If weight and bulk are objects, a foot or two of the straw can be cut from each sheaf or bundle of corn before it is taken to the stacks. The ears should be put inside, or half the corn will be taken by small birds; and the bottom of the stack should stand at least a foot from the ground. I use as food in winter peas, beans, barley, buckwheat, wheat, and a few oats, and many other little delicacies, such as boiled potatoes, ground artichokes, decayed apples, damaged raisins, &c.; and, with all these dainties, they will stray twice in the year—when the acorns fall, and at or just before breeding-time."

The following most complete series of suggestions on feeding pheasants in coverts is from the pen of Mr. James Barnes, of Exmouth. It is specially valuable as giving practical directions for the formation of catchpools for water, without which no amount of feeding will keep pheasants from straying in dry weather; and it also contains suggestions for the formation of huts, which are worthy of the careful consideration of every preserver on a large scale. Mr. Barnes writes: "Pheasants are well-known to require assistance with food of some kind in winter to keep them in good condition, and to have a propensity to ramble away and expose themselves to the depredations of trespassers. Buckwheat should be sown adjacent to their coverts, cut when ripe and intermixed with barley, also in straw, and placed in little stacks in or near their coverts, and spread or shaken about at intervals throughout the winter. What is still better to my mind, is to place their food in huts. A pheasant hut is an open shed, with the roof fixed on four posts, with a pole all round for rafter plate, the rafters of rough poles tied on with withies, thatched first with long faggots tied up with three or four withies of brushwood with all the leaves on, and allowed to hang down or over the rafter plate two feet or thereabouts. The thatch used should be small brushwood, reeds, or straw. An open trellis floor of poles should be raised two feet from the ground, and on this the corn in straw should be laid for the pheasants to help themselves. In these huts the pheasants find shelter, comfort, and cover in rough, wintry, and severe weather. Care should be taken to have plenty of dry dust on the floor underneath for the pheasants to bask in. This is a most essential provision—quite as much so for pheasants as for our poultry—for it is quite as natural for them to dust to clean themselves. It is a fact within easy observation how the pheasant searches out the base of an old dry, dusty pollard tree or hedge bank to bask in the dust. Besides, every grain of corn that falls through the open feeding floor is searched for and found in this dust. Underneath and on the dusty floor is a safe and convenient place, sheltered from severe frost, &c., to receive any other kind of food, such as refuse potatoes, Jerusalem artichokes, mangolds, swede turnips, cabbage, Spanish chesnuts, acorns, beechnuts, a few raisins, Indian corn, or anything else you wish the pheasants to have. Such changes of food cast about their feeding sheds are sure to secure them keeping pretty well to covert, particularly if they have water at hand. I have seen large expenditures for well digging or for the conveyance of water by ram and pipes from some stream at a distance; but the best and simplest plan to keep up a general supply of water for the season the pheasant is in covert, is certainly the shallow catch-pool system. In my humble opinion, it is the most natural, convenient, and inexpensive plan of all I have seen or had anything to do with in my time. I will explain what I mean by catchpools: Choose any little slope or valley in high and dry coverts where some command may be had of the surrounding surface water after rain; scoop out a hole in the earth's surface in the shape of a spoon or bowl, sloping gradually all round to the centre and deepest part, which need not be deeper than from eighteen inches to three feet, according to width and length; the edges, to admit the water running into it freely, must be kept a little under the earth's natural surface. Then puddle the whole of its face with six inches of well-wrought clay, paving it with bricks laid flat, and giving it all over a little coat of Portland cement. Thus you have a first-class and lasting catchpit to hold water most of the year, indeed, the whole season. Pheasants are expected to remain in covert for food and safety from September to February, and then there is certainly always plenty of water. After February the pheasant likes to go further away, and, soon after the gun is withdrawn, is pretty sure to get distributed about in search of insects and various roots. Pheasants rove about quietly during their breeding season, but little is seen or heard of them after April till corn harvest, as they live a quiet, secluded life through summer. I have made catchpools by casing them only with puddled clay. One disadvantage of this is, in a long dry time the water gets low, and the clay sides becoming exposed, contract, crack, and allow the water to run to waste if they are not looked to when rain does come. There is also another way in which I have had catchpools made where natural gravel abounds, namely, to make it into concrete and case the bottom and sides with this only. It answers well, and saves the labour and expense of getting bricks from a distance. Every feeder knows that dry barley and buckwheat in sheaf, and stacked in the vicinity of the preserves, and some pulled out and shaken about occasionally, with a change of maize, will keep the pheasants in good

FEEDING IN COVERTS. 57

condition; but it does not occur to everyone that a good supply of water near their feeding ground has a considerable influence on their habits. After feeding heartily on dry food, they will stray for water if there be none handy, and will stay away afterwards till hungry again, thus running the risk of being shot during their wanderings. To keep pheasants in their own coverts, take means of making them fond of them, even though there be no water near I have found Jerusalem artichokes the best means of attraction. They are so fond of these tubers that they will hunt them by sight or smell from any obscure corner. Give them also potatoes (small and large), mangold wurtzel, carrots, white-hearted cabbage, and savoys, all of which they will readily eat, and which not only prevent their straying for water, but afford a change of food that is genial and natural to their taste and well-doing, besides economising their dry corn food. Where the coverts abound with acorns, beechmast, Spanish chesnuts, and groundnuts, the pheasant requires but little feeding till the middle of December."

The planting of Jerusalem artichokes on waste spots and coverts will be found to be an exceeding advantageous mode of feeding pheasants and preventing their straying from their own coverts. When once established, these plants readily reproduce themselves and afford a continual amount of food to the birds. For preventing pheasants straying, the use of raisins scattered in the coverts is particularly advantageous. They will attract birds even from distant coverts to so great an extent that the owners of these latter may have to employ them in their own defence. So attractive are raisins to pheasants that the birds are not unfrequently captured by poachers by means of a fish hook baited with a raisin and suspended about the height of a running bird's head from the ground.

CHAPTER V.

MANAGEMENT OF PHEASANTS IN PRESERVES (CONTINUED).

REARING AND PROTECTION.

WITH regard to the rearing of pheasants in preserves but little need be said; the less they are interfered with the better. No good can possibly come from disturbing the sitting hens, but, on the other hand, a great amount of mischief may accrue. When leaving the nest quietly in order to seek food, the hen does so in such a manner as not to attract the attention of the numerous enemies, as crows, magpies, jays, &c., that are on the watch to discover and devour her eggs; but driven off by the prying intrusion of a visitor, she departs without caution, and makes known the situation of her concealed nest. The only circumstance warranting any interference with the nests of the wild birds is the occurrence of a greater number of eggs than the parent hen is capable of rearing as young birds, should the whole of them be hatched. A hen pheasant is rarely seen with more than six or seven young, at least when they have arrived at any size; and as she not unfrequently lays a larger number of eggs, it is an advantageous plan to remove all beyond eight or nine for the purpose of hatching them under common farmyard hens. Mr. J. Baily, in his "Pheasants and Pheasantries," says that if "a keeper knows of forty nests, seven eggs may be safely spared from each; this will give two hundred and eighty eggs for tame rearing"; but such a degree of prolificacy in wild pheasants is a higher average than has ever come under my notice.

Another point of very considerable importance with regard to the breeding of pheasants in preserves is the number of cocks that should be left in the spring in proportion to the number of hens. There is no doubt whatever that in a state of nature pheasants are polygamous, the stronger males driving away the weaker, and taking possession of several hens to constitute their seraglios; hence the custom to shoot down most of the cocks, and leave all the hens, even the oldest, to breed. It is probable that this procedure is frequently carried too far, and in confirmation of this view I have much pleasure in quoting Mr. J. D. Dougall, who, in his "Shooting Simplified," says: "It is customary to shoot cock pheasants only, and to impose a fine upon the sportsmen who break this rule, the money being escheated to the head keeper, or applied to defray the expenses of a dinner at the end of the season, when shootings are rented by a party of gentlemen. This rule is very frequently overstretched. It should not be forgotten that the desired end may be frustrated by having too many hens, as well as by having too few, and in whatever way the disproportion of sexes is caused, the result—reduction in increase—is the same. If the cocks are continually killed down, few male birds will arrive at that complete maturity so essential to producing a healthy stock. On the other hand, if the hens are continually spared, they will not only grow out of proportion to the number of cocks, but the aged hens will beat off the two and three year old birds. Very old hens should certainly be destroyed. The most prolific are the two and three year old birds."

A correspondent who supports this view writes: "It is very certain that in many instances too few cocks are frequently left in preserved coverts at the end of the season;

it is also notorious that in the neighbourhood of many preserves a nide of above fourteen birds (and I have known eighteen) is not unfrequently produced from an outlying cock and hen occupying some detached covert, and yields the best birds of the season when the 1st of October arrives. With respect to the proportion of cocks to be left much may be written about it, depending upon all circumstances connected with the ground under the entire control of the individual seeking to preserve a given stock of pheasants. In all cases, in my opinion, too much forbearance is shown to hens early in the season, and much too little towards cocks at the end. The safe plan, in all cases, is to adapt one or two small coverts, as much in the centre of your ground as possible, as your feeding places for your stock birds, and before the middle of December the exact number of birds which by judicious management you have collected there may be ascertained by a few days' careful observation. With attention and the greatest forbearance towards these (no old cocks being left among their number), you may kill freely elsewhere, and insure to your friends and yourself plenty of sport the following season from them and their progeny."

With regard to the exact proportion of sexes left in the coverts, it is difficult to arrive at a satisfactory conclusion. One writer states: "It would be to the advantage of preservers of pheasants if they would, before it is too late, refrain from shooting the cock birds too close, as most game preservers, I presume, wish to have as good and numerous a stock of pheasants as they can for breeding; and the reason why so many are disappointed in this respect is for want of more cock birds. There should be left at least one cock for every three hens, as eggs then would be more plentiful, the chicks stronger, and better able to contend with a wet season and the numerous enemies they have to battle with."

The frequent occurrence of old barren hens that have assumed either wholly or in part the plumage of the male is a proof, if one were wanting, that in many coverts the old worn out hens are left longer than is desirable or profitable.

The chapters on the Management of Pheasants in Preserves would be very incomplete without the consideration of the best means of protecting them against their numerous enemies. The chief four-footed depredators are cats, foxes, hedgehogs, and polecats. Their other enemies are feathered and unfeathered. Amongst the former may be included crows, magpies, and jays, which are great destroyers of eggs. But the unfeathered bipeds, known as poachers, are perhaps the most destructive. By far the greater number of pheasants purloined by the poacher are shot at night; this destruction may be prevented in great part, without the necessity for night watching, by having suitable coverts, as has been already fully explained in the preceding chapter. Where larches and other trees with exposed horizontal branches abound, recourse should be had to mock pheasants, which are excessively annoying to poachers, as they cause them to expend ammunition uselessly and alarm the neighbouring keepers, without any profitable result. Mock pheasants, quite incapable of being distinguished from the real birds at night, may be made of hay bands, rushes, or fern, bound with tarred twine or wire on a stick about two feet long. Capt. Darwin, in his "Game Preserver's Manual," writing of mock pheasants, states "they are very easily made, but their situations should be often varied. Some keepers make them of board cut into the shape of a pheasant. These are of little use, for a poacher gets under them and sees at once what they are. Others make the body of wood, roughly turned in a lathe, and nail a strip of wood on it for a tail, or with real tail feathers stuck in. The best mode of making mock pheasants after all is as follows: Get a bunch of long hay and roll it round a stick till it is the size of a pheasant's body, leaving enough for a tail; wrap it with thin copper wire down to the end of the tail; cut a peg about six inches long and as thick as a lead-pencil; wind a bit of hay round the end to make a head, and run the peg into the body. Tie these imitations on the branches of larch trees here and there. Pheasants prefer this kind of tree to others, in consequence of the boughs coming out straight, and so allowing them a flat surface to sit on. In woods where there are no foxes, and where the ground vermin has been well killed down, it is a good plan (especially if you think it a likely night for poachers) to unroost the pheasants in the evening. They will not fly up again that night. If you begin by unroosting the pheasants when they are young, and have only flown up a few nights, they will take to roosting on the ground altogether, and never fly up at all. Pheasants that have not been accustomed to be driven down at all are made rather shy by the frequent repetition of this performance, and it may drive them away. They are very easily frightened. If you begin shooting rabbits, &c., they will take the alarm. They can't stand guns going off constantly in the coverts where they are."

Imitation pheasants thus made will only last a single season; should anything more permanent be desired, recourse must be had to those made of wood, which may be cheaply and efficiently constructed on the following plan, the suggestion of a correspondent, who states: "Six years since I had

a number of wooden ones made and set up, and hundreds remain to this day. The manufacture was simple. Take a fir pole, saw it through at an angle of 45°; this cut, when rounded off, forms the breast of the bird; a cut at 22½° forms the tail-end. So, by making alternate cuts at 45° and 22½°, you may cut up the pole without waste, as shown in the plan sent herewith. A lath cut through in like manner at a very acute angle forms a capital tail, which should be put on, but nearer the perpendicular than shown in the engraving, as pheasants roost with the tail hanging nearly straight down, the head is easily made out of the upper end of the pole, where too small for the body. Daub over with some oil paint (burnt umber), bore a hole through the body for the nail, and nail on the tree with a chisel-ended nail, that you may not split the branch. What the cost is you may judge, as a 12ft. pole costs fourpence, or less. Place them pretty thick where

pheasants roost. By boring a hole 1in. diameter from the underneath to within half an inch of the back, they will, if placed on a nail, move with the wind. My experience of them is that the deception is perfect enough, as they are difficult to distinguish from a pheasant, even in daylight.

Whatever kind of mock pheasant is employed, they should not be placed too near public roads or footpaths, and in those cases in which they are liable to observation during the day, they should be moved frequently."

Alarm guns set in coverts with wires leading in different directions are most valuable as alarming poachers, and indicating the locality in which they are pursuing their depredations. One of the best, and certainly the cheapest, alarm guns with which I am acquainted, is that devised by Captain Darwin, and described in his useful manual on Game Preserving, which has been too long out of print. The author writes: "I have constructed an alarm gun which combines the desiderata of cheapness and simplicity more completely than any I have yet seen. I do not lay claim to the invention of this gun, but I certainly find I can adopt materials in its construction that will come to a tenth part of the money usually charged; in fact, any tolerable mechanic ought to make it in an hour. It is formed as follows: get a piece of iron gas-pipe, three inches long and three-quarters bore. At the threaded end make a plug of iron a quarter of an inch thick, and tapped in the centre for a nipple. Drive this plug into the barrel, and *braze it*. The nipple is then screwed in. Then get a corresponding piece of the gas-pipe, from two feet six inches to three feet long, also threaded at the end. Screw the collar (that always goes along with this sort of gas-pipe) on to the long piece as tight as it will go. The gun is now complete with the exception of the hammer, which is a piece of round iron about a foot long, and slipping easily down the barrel. To set the gun you must tie the long barrel fast to the stem of a tree in the plantation, with the short barrel downwards. Unscrew the latter and load it with a couple of charges of powder, and put on the cap, which you should cover with some beeswax and suet mixed. Then screw the short barrel into the long one. Drill a small hole through the loose piece of iron about four inches from one end, and put it in the barrel with a nail or peg in the small hole, and a string from the nail going down the side of the tree in the direction you may choose. Mind and not have the wire so low that a dog can let it off. When the wire is touched it draws the nail, and the hammer, falling down on the barrel, lets the cap off. Being fastened up in a tree, and close to the stem, it can catch the eye of no one, and merely has to be shifted occasionally, though of course there is no need to do this until after it has been fired. After all, nothing daunts poachers so much as pit-falls made in the woods. They should be about seven feet deep, and made with the sides slanting, so that the chamber is larger at the bottom than at the top. Unless boarded all round, the soil will fall in. The opening should be four feet square, and be covered with sticks and sods, or anything resembling the surrounding ground. Poachers are very shy of venturing into woods where you have these pit-falls."

Alarm guns discharging wooden or other plugs upwards

or horizontally should never be used, as danger to human life always accompanies their employment. It is almost unnecessary to remark that alarm guns of various forms can be purchased at any gunmakers.

The destruction effected in preserves during the nesting season by crows, jackdaws, magpies, jays, and other egg-eating birds, is well known, and can only be remedied by the trapping or shooting of the culprits. The question as to the influence of the rook in pheasant coverts is one of those respecting which there is much to be said on both sides. The rook is so very valuable an ally to the agriculturist, by destroying an enormous number of grubs, wire worms, &c., that its case claims our most attentive consideration. In reply to the accusation that rooks occasionally destroy the eggs of the pheasant, Mr. James Barnes writes: " According to my own observations of above fifty years, the rook will eat eggs if placed about in open country pastures, &c., but I believe never goes on foraging excursions for eggs or young game, as the carrion crow does. Rooks will not only knock eggs to pieces openly placed in sight of their feeding grounds, but they will also, in hard frosty weather, devour many other things, such as slaughter-house garbage, or dead poultry, game, or fish that may lie about decomposing within their reach. My own observation is, that the rook is a real friend to the pheasant, and provides it with a deal of food at an acceptable season. In the years 1816 and 1817, I went with others to see the young rooks shot in Lord Middleton's park, Peper Harrow, Godalming, Surrey. The trees were high in an inclosure, but not at that time very thick on the ground, for there was some scrubby undergrowth and a rare crop of rank weeds—the open spaces were splashed as if whitewashed, as the undergrowth of all rookeries is during the first two or three weeks of May. Amongst this undergrowth there were two or three pheasants' nests, protected with boughs; and strict orders were given that no one should disturb the pheasants' nests. I thought but little of this at the time; but afterwards I observed that where pheasants were preserved near a rookery, pheasants were to be seen there through March, April, and May. I did not observe the real cause of their foraging and running about the rookeries till about 1844, when I saw a cock pheasant pick up a piece of potato on a gravel walk, and run away with it into the shrubbery, and remembered that I had often seen pieces of potato lying about, and had seen the rooks drop them and their pellets likewise. The latter were frequently full of half-digested grains, as if dropped through fright. I had seen from the middle of February to the middle of May bushels of pellets underneath the trees scratched over by the pheasants—of course for the food to be found therein ; and there were always pheasants' nests close at hand, even in or under the rookery. Where the potato is much cultivated, as in South Devon, a good many small potatoes would be turned up in ploughing the land, which the rook and jackdaw seemed to claim as their perquisites and carry off home. I have seen five or six fall of a morning on walking under the trees, but the birds never came down to pick one up. I have seen fall large brown grubs, the fern beetle, whole ears and loose grains of corn, pellets or quids half chewed or sucked over, and have seen the pheasants run and pick them up. There is fine living in variety for pheasants under a rookery, provided neither party is disturbed by strangers. Respecting the rooks' pellets, from the middle of February to the middle of March, in a corn-growing district, while the spring corn is sowing the rook hurries over the new-sown land, and picks up all stray grains that comes under his observation, as well as worms, grubs, slugs, bits of potatoes, pieces of half-decayed scales of oyster shells, little pieces of lime, sand, and gravel— all together hoarded under the lower mandible, which looks like a big full pouch as he arrives home to his mate in charge of the nest. Here his load is delivered to the mate, who, with great ado, chews it over, and ejects the pellet or quid in due course. This business is continued till late at night. Many times, passing under the trees at various hours, from 10 p.m.

till 3 a.m., I have heard the pellet drop, and have had them fall on my head and shoulders, and picked them up by the light of the moon or lantern. The rook's excreta are at this time pretty solid. As the month of March is nearly ended this alters; and in April, when the corn is sprouted and growing, the ejecta are like sloppy mud, and contain the husks of a few grains of corn, wings of beetles, pieces of snail shells, lime, and grit. From this time till June no pellets or quids are to be seen ; the droppings are loose, and like whitewash over the vegetation underneath. Insect food gets so various and abundant that they and their broods seem to entirely subsist on it for six or eight weeks, and the young thrive and grow fat wonderfully quick in showery, growing weather of April and beginning of May. The young that are spared from the gunners, as soon as they can fly, are enticed away early in the morning by their parents, at first by short flights, to the fields then preparing for turnip sowing, or the pasture that produces cockchafers, fern beetles, and other insects, and for a few nights roost on trees near their work. After they get strong on the wing, and good flyers, they all come back to their native home, the rookery. As soon as a field of early podded peas is pretty full, the rook, if not looked after, will take toll; also of wheat or barley they will certainly, if an opportunity is afforded them, filch a portion, particularly such as is near trees or has been laid by wind or wet. Then, again, commences the real pellet-ejecting season. The rook then hurries to the unguarded field to filch corn, which he stores in his pouch as quickly as possible, picking up also on the pasture and turnip fields, &c., quantities of grubs, snails, slugs, beetles, earwigs, grasshoppers, crickets, fern flies, various other insects, and their larvæ. It is truly astonishing to see, as I have done for years, on examining those ejected pellets, what variety at times they contain—besides remains of every kind of creeping, running, or flying insect that may chance to come in their way, in the season of ripening of seeds on the pastures a number of grass and weed seeds, the husk of corn—wheat and barley—many kinds of weed and coarse grass seeds. After harvest and gleaning season is over, no more pellets are to be seen. In the wheat-sowing season they filch some loose grains and dig out the young plants, and, through its being wet at this season, and collected with much dirt, the food is ejected in a loose manner like mud. However, in all my long experience, I never saw under the trees an eggshell of any game, poultry, or other birds, except the shells of their own which had been hatched out, or tumbled out by stormy winds. I have, however, yearly seen a pair of carrion crows attend to the early rooks' nest, and carry off the new-laid eggs, as they did also with pheasants' eggs, the shells of which I have found lying about by scores. It is a curious fact that, numerous as the rooks are, they are such cowards as to allow the crow to rob them, and only fly round and round, cawing, while the robbery is going on."

I have known many cases where pheasants have sat, and reared their young safely almost immediately under a rookery. On the other hand, there is no doubt but that in seasons of scarcity, when very hard pressed for food, rooks will destroy pheasants' eggs.

Colonel J. Whyte, Newtown Manor, Sligo, in reply to Mr. Barnes, writes as follows : " There appears some doubt whether rooks suck pheasants' eggs, or whether the carrion crow is not the real depredator. Perhaps what follows may set the question at rest. About four years since, Lord Clonbrock asked me if I had ever known rooks eat the eggs of pheasants. My idea was that they might do so occasionally, but not as a custom. His lordship replied: ' The rooks about me have within the last year or two taken to hunt up and destroy the eggs as regularly as if they were so many magpies. I did not believe my keeper at first, but, going myself to look out, I saw them regularly beating up and down a piece of rough ground where the pheasants nest, and when they found one they would rise up a few yards in the air and then pounce down on it.' Lord Dunsandle's place is within fifteen or sixteen miles of Lord Clonbrock; there are three rookeries in it, and the first question I asked the keeper on my arrival there to shoot was, 'Do the rooks suck or damage the pheasants' eggs?' The answer was, 'No;' nor did they do so till this year. But about a week ago I received from Lord Dunsandle a letter, in which he said, 'This year the rooks have taken to destroying my pheasants' eggs, and the mischief they have done is incredible; the fields are strewn with broken eggs.' It would therefore appear that not only do rooks destroy eggs, but that they take to it in a sudden and unaccountable manner. There can be no question here about the work being done by carrion crows, for the only carrion crow in Ireland is the Royston or hooded crow.* The reason that Mr. Barnes had no shells under the trees in the rookery is, that the rook breaks and eats the eggs on the spot. Jackdaws will eat eggs whenever they can find them, and my keeper assures me that a short time since he saw one take a little rabbit up in his claws several yards, and then drop it on his approach. This colony of jackdaws is situated in some high cliffs, and is increasing in numbers every year."

Mr. Leno, a very extensive pheasant breeder, states the case still more forcibly:—"My experience is, that rooks will destroy pheasants' eggs whenever they happen to find them out. In one week a rook came twice and settled down in my pheasantry, and took an egg away each time: and where rooks abound, if perchance a pheasant's or partridge's nest is left by the mowers, the rooks may be seen crowding around the patch of grass left for shelter, and the eggs are finished in quick time. It is useless to leave a nest exposed in the neighbourhood of rooks, as they are sure to eat them."

Mr. Harman, of Riverstown, co. Sligo, writes: "I am unwilling to bring in a case against that useful bird the rook,

---

* It is now ascertained that the Carrion and the Royston crows are merely varieties of one and the same species, and that they breed together with great freedom. Both varieties occur in Ireland.

but I can confirm the destruction of pheasants' eggs. A few years ago, in a dry spring, with a north-east wind for many weeks, when the rooks could not bore for their accustomed food, about one hundred and fifty pheasants' eggs—*i.e.*, the shells—were found under the rookery near the house, having been taken by the rooks to feed their young, other food failing them. I have caught them when baiting traps with eggs for magpies ; but still I consider the rook (barring these serious misdemeanours) a most useful bird."

Mr. J. E. Harting informs me that on one occasion, in the month of April, about the 14th or 15th, he saw a rook in the act of carrying off a pheasant's egg from a copse. The bird was carrying the egg upon the point of the bill, and on being fired at he dropped it, and when picked up it was found to be empty, although still wet inside. There was a large and irregularly shaped hole towards the larger end. On the very ground where this occurred, my informant had heard the keeper say that he had on more than one occasion shot rooks in the act of carrying off pheasants' eggs.

The balance of the evidence for and against the rook in respect of its conduct regarding the eggs of pheasants, appears to show that, saving in seasons of an exceptional character, or in cases where the eggs are left exposed by mowing, the influence of the bird is not seriously antagonistic to the rearing of pheasants ; but when hard pressed for food, rooks will even destroy the young birds. A correspondent writes as follows :—" On June 13 my keeper observed about half a dozen rooks engaged amongst the coops of young pheasants, and, suspecting their object, drove them off. The next morning, having fed and watered the young birds, he went to his cottage, and, looking out about six o'clock, saw a strong detachment of rooks from a neighbouring colony in great excitement amongst the coops. He ran down, a distance of two hundred yards, as fast as possible, but before he arrived they had succeeded in killing, and for the most part carrying off, from forty to fifty birds, two or three weeks old. As he came amongst them they flew up in all directions, their beaks full of the spoil. The dead birds not carried away had all of their heads pulled off, and most of their legs and wings torn from the body. I have long known that rooks destroy partridges' nests and eat the eggs when short of other food, but have never known a raid of this description. I attribute it to the excessive drought, which has so starved the birds by depriving them of their natural insect food that they are driven to depredation. It will be necessary to be on guard for some time; bad habits once acquired (as with man-eating tigers) may last even more than one season. Probably the half-dozen rooks first seen amongst the coops tasted two or three, and, finding them eatable, brought their friends in numbers the next morning."

The Moorhen, Waterhen, or common Gallinule, is occasionally destructive to young pheasants. Mr. Gould recounted the evidence in "The Birds of Great Britain," and Mr. H. J. Partridge, of Hockham Hall, Thetford, writing to the *Zoologist*, stated that "At the beginning of July, the keeper having lost several pheasants about three weeks old from a copse, and having set traps in vain for winged and four-footed vermin, determined to keep watch for the aggressor, when, after some time, a Moorhen was seen walking about near the copse; the keeper, supposing it only came to eat the young pheasants' food, did not shoot it, until he saw the Moorhen strike a young pheasant, which it killed immediately, and devoured, except the leg and wing bones. The remains agreed exactly with eight found before."

Lord Lilford, writing in "Dresser's Birds of Europe," says: "I look upon the Waterhen as an enemy to the game-preserver, not only from the quantity of pheasant food which it devours, but from the fact that it will attack, kill, and eat young birds of all sorts. The bird is a great favourite of mine, and I should be sorry to encourage its destruction, but I am persuaded that it is a dangerous neighbour to young game birds"; and in his "Birds of Northamptonshire," he adds, "We cannot acquit them of the charge of a very pugnacious and destructive tendency amongst their own and other species of birds, and they are most certainly bad neighbours for young pheasants and partridges, as they not only consume a good deal of the food intended for game birds, but will now and then capture and devour the birds themselves."

The common Kestrel, or Windhover, so well known as a destroyer of field mice and rats, has also been accused of attacking young pheasants. Mr. J. H. Gurney, of Northrepps, one of the highest authorities on accipitrine birds, writes as follows:—"Mr. Stevenson, in his article on the Kestrel in the 'Birds of Norfolk,' remarks: 'That some kestrels carry off young partridges as well as other small birds during the nesting season, is too well authenticated as a fact for even their warmest advocates to gainsay.' For many years I have endeavoured to collect reliable information on this point, and I am convinced of the correctness of Mr. Stevenson's opinion above quoted; but there is this difference between the sparrowhawk and the kestrel in their habits of preying on young partridges and pheasants—viz., that the kestrel only destroys them when very young, and the sparrowhawk continues to attack them long after they have grown too large to be prey for the kestrel. To particularise two instances: Many years ago, a very young partridge was brought to me which had been taken out of a kestrel's nest at Easton, in Norfolk; and a gamekeeper in this parish, who is as trustworthy an observer of such matters as any man I know, saw a hen kestrel take up a very young pheasant in its talons and rise with it about eight feet from the ground; my informant then fired at the depredator with a small pistol, when it dropped its prey, which, though somewhat injured, ultimately recovered; and an instance of a young pheasant found in the nest of a kestrel was recorded in *The Field* of May 13, 1868."

Mr. Booth, in his "Rough Notes," has carefully investigated the accusations against the kestrel, and he maintains that it is one of our most useful birds, and a decided ally to the game preserver, more especially as a destroyer of rats, of which it kills large numbers. He says he has never known the kestrel to carry off young broods of either pheasants or partridges, but that the damage done by the sparrowhawk is often attributed to the Kestrel.

The pheasant, from nesting on the ground, is peculiarly exposed to the attacks of four-footed or ground vermin, and the escape of any of the sitting birds and their eggs from foxes, polecats, hedgehogs, &c., appears at first sight almost impossible. This escape is attributed by many, possibly by the majority, of sportsmen to the alleged fact that in the birds when sitting the scent which is given out by the animal at other times is suppressed; in proof of this statement is adduced the fact that dogs, even those of the keenest powers of smell, will pass within a few feet, or even a less distance, of a sitting pheasant without evincing the slightest cognizance of her proximity provided she is concealed from sight. By others this circumstance is denied, they reason *à priori* that it is impossible for an animal to suppress the secretions and exhalations natural to it—secretion not being a voluntary act. I believe, however, that the peculiar specific odour of the bird is suppressed during incubation, not, however, as a voluntary act, but in a manner which is capable of being accounted for physiologically. The suppression of the scent during incubation is necessary to the safety of the birds, and essential to the continuance of the species. I believe this suppression is due to what may be termed vicarious secretion. In other words, the odoriferous particles which are usually exhaled by the skin are, during such time as the bird is sitting, excreted into the intestinal canal, most probably into the cæcum or the cloaca. The proof of this is accessible to every one; the excrement of a common fowl or pheasant, when the bird is not sitting, has, when first discharged, no odour akin to the smell of the bird itself. On the other hand, the excrement of a sitting hen has a most remarkable odour of the fowl, but highly intensified. We are all acquainted with this smell as increased by heat during roasting; and practical poultry keepers must have remarked that the excrement discharged by a hen on leaving the nest has an odour totally unlike that discharged at any other time, involuntarily recalling the smell of a roasted fowl, highly and disagreeably intensified. I believe the explanation of the whole matter to be as follows : the suppression of the natural scent is essential to the safety of the bird during incubation; that at such time vicarious secretion of the odoriferous particles takes place into the intestinal canal, so that the bird becomes scentless, and in this manner her safety and that of her eggs is secured. This explanation would probably apply equally to partridges and other birds nesting on the ground.

The absence of scent in the sitting pheasant is most probably the explanation of the fact that foxes and pheasants are capable of being reared in the same preserves; at the same time the keepers are usually desirous of making assurance doubly sure, by scaring the foxes from the neighbourhood of the nests by some strong and offensive substance. A very practical gamekeeper writes as follows:— "If any keeper will find his nests and sprinkle a little gas tar anywhere about them, he will find the foxes will not take the birds. I should, as a keeper, find every nest possible, and dress the bushes, stumps of trees, &c., near the place of such nest, and then keep away entirely till I thought the bird had hatched, as constantly haunting a bird's nest is the most foolish thing that can be. When such nests are once found and dressed, let the keeper look out and trap all kinds of vermin, such as the cat, stoat, fitchet, weasel, hedgehog, or rat, or magpie, jay, hawk, crow, rook, or jackdaw. These are all enemies to the birds, as well as the fox. I am satisfied, as a gamekeeper, that with good vermin trapping, dressing near the nests, and good bushing and pegging of land, anyone will have plenty of game, and may still keep plenty of foxes."

Another equally efficacious plan, the value of which has

been repeatedly proved, is to fill a number of phials with the so-called "oil of animal" (also known as oil of hartshorn and Dippel's oil), and suspend them uncorked to sticks about eighteen inches long, and stick two or three round each nest, about a foot from it. The smell of the oil will keep the foxes from approaching.

In the vicinity of dwellings, there is no more dangerous enemy to pheasants than the common cat. Captain Darwin, in his "Game Preserver's Manual," writes as follows:—
"There is no species of vermin more destructive to game.than the domestic cat. People not aware of her predatory habits would never for a moment suppose that the household favourite that appears to be dozing so innocently by the fire is most probably under the influence of fatigue caused by a hard night's hunting in the plantations. How different also in her manner is a cat when at home and when detected prowling after the game. In the first of the two cases she is tame and accessible to any little attentions; in the latter she seems to know she is doing wrong, and scampers off home as hard as she can go. Luckily there is no animal more easily taken in a trap, if common care be used in setting."

Laying poisoned meat is now illegal, and the sale of arsenic to private persons interdicted by statute; nevertheless I would caution any one against the use of that drug, as the employment of it is attended with much cruelty, as it is immediately rejected by vomiting, but not before it has laid the foundation of a violent and painful inflammation of the stomach, from which the animal suffers for weeks, but rarely dies. If it is absolutely necessary to use poison for cats, a little carbonate of baryta, mixed up with the soft roe of a red herring, is the most certain and speedy that can be employed, but a good keeper should know how to keep his preserves clear of vermin without the aid of poison.

Hedgehogs are undoubtedly destructive to eggs as well as to the young birds, and should be trapped in coverts in which pheasants are reared.

Among the other enemies to young pheasants that attack them occasionally may be mentioned adders, and even tame farmyard ducks that have gained access to the coops.

The following vignette shows the extraordinary manner in which wounded and malformed pheasants adapt themselves to new conditions of life. It represents most accurately the head of a ring-necked pheasant that was killed by Mr. Godwin on Lord Torrington's estate. The bird was in very fair condition, weighing 2lb. 5oz., and had thirty-three beech nuts in its crop. Both mandibles had been cut off in front of the nostrils, most probably by a strong steel trap, the tongue, however, had escaped, and protruded from the mouth. It is difficult to imagine that the bird had the power of taking up small grains, and it is not surprising that it fed mainly on beech nuts, which it could readily take into its mouth.

CHAPTER VI.

MANAGEMENT OF PHEASANTS IN CONFINEMENT.

FORMATION OF PENS AND AVIARIES.

HAVING treated of pheasants as wild birds, their rearing and management in enclosed pens and aviaries have next to be considered. When pheasants are bred for turning out into the coverts, and not as merely ornamental aviary birds, the system of movable enclosures, constructed of rough hurdles, will be found far superior to any more elaborate contrivances, for, when the breeding birds are kept in the same place year after year, the ground becomes, in spite of all the care that may be bestowed on it, foul and tainted, disease breaks out even amongst the old birds, and the successful rearing of young ones is hopeless.

The pens should be situated in a dry situation, sandy or chalky if possible, but any soil not retentive of wet will answer. If the surface is sloping it is to be preferred, as the rain is less likely to render the ground permanently damp. Although cold is not injurious to the mature birds, and they require no special shelter, the south side of a hill or rising ground is to be chosen in preference, as the young stock are delicate. Common wattled hurdles, made seven feet long, and set up on end, make as good pens as can be desired; they should be supported by posts or fir poles driven firmly into the ground, with a horizontal pole at the top, to which the hurdles are bound by tarred cord, or, still better, very stout flexible binding wire, which should also be used to secure them together at top and bottom. The posts should be inside the pen, as better calculated to resist any pressure from without.

The hurdles should rest on the ground without any opening below, and if they are sunk three or four inches below the surface, the pens will be more secure against dogs and foxes or any animals likely to scratch their way under. The size of these pens should be as large as convenient; for a cock and three to five hens—the utmost number that should be placed together—as many hurdles should be employed as will form a pen twenty-five to thirty-five feet square, the smaller containing 625 square or superficial feet of surface; the larger, which will require less than half as many more hurdles, containing nearly double the interior space, namely, 1225 square feet. If the birds are full winged, these enclosures must be netted over at the top; for this purpose old tanned herring netting, which can be bought very cheaply, will be found much better than wire-work, as the pheasants are apt, when frightened, to fly up against the top of the enclosure, and, if it be of wire, to break their necks or seriously injure themselves. Should netting be employed, several upright poles, with cross pieces at the top, are required to be placed at equal distances to support the netting, and prevent it hanging down into the interior of the pen. A much better plan is to leave the pen quite open at the top, and to clip one of the wings of each bird, cutting off twelve or fourteen of the flight feathers close but not into the quills. When the birds cannot fly they become much tamer, are more productive, and are not so apt to injure themselves by dashing about wildly, especially if there be, as is desirable, brushwood cover or faggots in the pen, under which they can run and conceal themselves. Some persons are in the habit of pinioning the birds by cutting off the last joint of the wing, thus removing permanently the ten primary quills, but the plan is not to be recommended, as the pinioned birds are quite incapable of taking due care of themselves when turned out into the open, and are liable to fall a prey to ground vermin.

As illustrative of the mode in which a large number of birds can be successfully kept in one locality, I will describe the arrangements which I saw at the pheasantries belonging to Mr. Leno, a very successful rearer. The birds are kept in runs enclosed by hurdles between six and seven feet high. These are formed of stout straight larch laths nailed to cross pieces of oak or other strong wood, and are fastened to stout posts securely driven into the ground. As the posts are capable of being easily withdrawn and replaced, there is no difficulty in moving the pens year after year—a most important consideration for the preservation of the health of the birds. Moreover, by employing a greater or smaller number of hurdles and posts, pens of any required size may be constructed, so as to accommodate a larger or smaller number of birds. On my visit, the runs had recently been shifted on to new ground, which consisted of young hazel coppice, which had been partly cleared. The surface was covered with the dead leaves of last year's growth and with short underwood, affording ample opportunity for the birds to amuse themselves by scratching for insects and by seeking food amongst the leaves. The amount of undergrowth afforded another important advantage, that the birds, on the entrance of a stranger, could run under shelter, and so conceal themselves, instead of dashing about wildly, as they would otherwise have done. No roof or shelter of any kind was afforded them, had such been erected the birds would only have used it for roosting upon, and not for sleeping under. In each pen was a horizontal pole, supported about four feet from the ground by a post at each end. Across this was laid a number of stout branches and long faggots, forming a kind of shelter to which the birds could have recourse, and under which the hens would occasionally lay; but the chief advantage it affords is that of a roosting-place, elevated from the ground, and so keeping the birds away from the cold damp soil during the night. The sloping arrangement of these branches is advantageous to the birds, as all of them have the flight feathers of one wing (not both) cut short; they are thus destitute of the power of flight, and consequently inclined branches, up which they can walk and down which they can descend without violence, are exceedingly useful. These runs, open as they are, afford all the shelter required, provided they are not placed on the north or east side of a hill or rising ground. Their advantage over permanent buildings is great; in the latter pheasants cannot be successfully reared, as the ground becomes tainted, epidemic disease breaks out, and the ground also becomes charged with the ova of the *Sclerostoma syngamus*, or gapeworm, which often causes great havoc amongst the young poults. Both of these evils may be in great measure avoided by shifting the runs as frequently as may be convenient. The runs may be made of any size, so as to accommodate one cock and three or four hens, or a larger number of birds. Care must be taken not to have them too small, as the birds when closely confined, often take to pecking one another's feathers—an evil which is occasionally carried on until the persecuted bird is killed. When runs are made small, the ground very rapidly becomes tainted, and the birds consequently diseased. The vigorous, healthy aspect of the numerous birds I saw at these pheasantries was evidently owing, in great part at least, to the large size of the inclosures, and the fresh ground on to which they are so frequently shifted. No nest-places are made or required; the hens generally drop their eggs about at random, and they should be looked for and collected at least twice a day. This is most important, as, if any eggs are chipped or broken the birds may acquire the bad habit of pecking them, which is quickly acquired by all others in the run, and will be found exceedingly difficult to eradicate. The food employed is good sound barley, with a certain proportion of buckwheat. This is varied by soft food consisting of meal, with which, at times, a small proportion of greaves is mixed to supply the place of the animal food the pheasants would obtain in a state of nature. Acorns are occasionally employed, but the birds prefer grain. The food is strewed broadcast on the ground; and it is needless to say that a constant supply of clean fresh water is provided for the birds. The young are hatched under common barnyard fowls, and are reared on custard, biscuit, meal, rice, and millet, with occasionally a little hempseed—ants' eggs, though exceedingly advantageous, not being found in the locality.

The arrangements recommended by Mr. F. Crook vary somewhat in detail from those described, but are equally practical and effective. He writes:—" An order should be given to the ordinary wattled-hurdle makers to make a given quantity of six feet by six feet open hurdles, with well-pointed ends, twenty-four of these hurdles, when placed in position, will make a convenient-sized run, thirty-six feet every way; but preparation must be made for a doorway, and for covering over the whole of the hurdles inside the run with one and a half inch wire netting round the sides, and string netting for the top. For the size run specified there must be four posts, made with four-way T piece tops, to carry the netting; the posts to be placed equi-distant from each other, to properly divide off the interior centre space; from each upright should branch out movable perches about eighteen inches long, at different heights from the ground. The next and most important point is the arrangement of nesting-places. At the most retired portion of the run faggots should be placed, in bundles of three or more, arranged conical fashion, or piled as soldiers do their arms, leaving a good space open at the bottom; but before setting the faggots in their places, the earth must be dug out six inches deep, and filled in with dry loose sand or fine dry mould, and then place the faggots over the sand. There should be as many of these nesting-places as the space will afford, taking care that sufficient space is left between each to admit of easy access by the birds and their keeper." Some writers recommend pens made of eight hurdles, each six feet long, giving a square of twelve feet in each side, and having an interior space of only 144 superficial feet; but these pens are too small for the health or comfort of the birds, that are far more apt to fall into the evil habits of egg eating and feather plucking than when confined in larger runs.

With regard to the food of the old birds in the pens, the more varied it is the better. Good sound grain, such as maize, barley, buckwheat, malt, tail wheat, and oats, &c., may all be used. But maize should be used sparingly, as it is too fattening for laying pheasants or hens. Mr. Baily recommends strongly an occasional feed of boiled potatoes, of which the birds are exceedingly fond. He writes:—"For bringing pheasants home, or for keeping them there, we know of nothing equal to boiled potatoes. Let them be boiled with the skins whole, and in that state taken to the place where they are to be used. Before they are put down, cut out of each skin a piece the size of a shilling, showing the meal within. Place them at moderate distances from each other, and the pheasants will follow them anywhere."

Rice and damaged currants and raisins are very well for an occasional change, but should be sparingly used. A few acorns may be given from time to time, but their use in excess is apt to prove injurious. Mr. J. Fairfax Muckley, of Audnam, writes on their employment as follows:—" Three seasons ago I laid in a stock of acorns, and instructed the feeder to give the pheasants a few every day. They preferred them to other food. In one week I had ten dead birds. They were fat and healthy in every respect, with the exception of inflammation of the intestines. My conclusion is, that if allowed to have free access to acorns they eat more than they should, and consequently many die. Keepers frequently depend too much upon acorns."

With regard to the employment of animal food, such as

horseflesh, greaves, &c., I believe its use, except in the very smallest quantity, to be exceedingly injurious; nor do I approve of the spiced condiments so strongly recommended by the makers. The bodies of dead domestic animals can, however, be most advantageously utilized by allowing them to become thoroughly fly-blown, and then burying them under about a foot of soil in the pens, where the maggots go through the regular stages of growth, after which they work their way to the surface in order to effect their change into chrysalids. They furnish an admirable supply of insect food for the birds, and give them constant occupation and exercise in scratching in the ground. Utilized in this manner, the bodies of dead fowls, or any small domestic animals, are perfectly inoffensive, and the result is most advantageous to the birds.

The employment of crushed bones, as a substitute for the varied animal substances the pheasant feeds upon when in a wild state, is highly advantageous. Mr. F. Crook writes:— "We have seen many instances of game being perfectly cured of both eating their eggs and plucking each other, by the continual practice of giving a portion of well-smashed bones every day. These remarks apply more specially to the home pheasantries, in consequence of the absence of the natural shell stuff they pick up when at liberty, but we would recommend some to be thrown about the feeding grounds of the preserves, as the highly nutritious nature of the elements of smashed fresh bones conduces remarkably to keep the birds together, particularly in very wet seasons, when the condition of the land renders it impossible for them to scratch about to the same extent." Should the aviary be situated on soil in which small stones are absent, these must be supplied; this is most conveniently done by throwing in some fresh gravel once or twice a week.

There is one point on which almost all the works treating on the management of pheasants are lamentably deficient, namely, enforcing the absolute necessity for a constant supply of fresh green vegetable food. The tender grasses in an aviary are soon eaten, and the birds, pining for fresh vegetable diet, become irritable, feverish, and take to plucking each other's feathers. To prevent this, cabbages, turnip leaves—still better, waste lettuces from the garden, when going to seed—should be supplied as fast as they are eaten; the smaller the pen the greater the necessity for this supply. The late Dr. Jerdon, the distinguished author of "The Birds of India," when visiting the pheasantries in the Zoological Gardens, said, in his emphatic manner, "You are not giving these birds enough vegetable food. Lettuce! Lettuce!! Lettuce!!!" From my long experience in breeding gallinaceous birds of different species, I can fully indorse his recommendation.

Should these cultivated vegetables be not readily obtained, a good supply of freshly cut turves, with abundance of young grass and plenty of clover, should be furnished daily.

Instead of placing a cock and three to five hens in a pen, as recommended, some persons advocate putting cut-winged hens only in enclosures open at the top, so that they may be visited by the wild males. Of necessity, this method can only be followed in the immediate vicinity of coverts well stocked with pheasants, and even under these conditions it is not always successful, the eggs frequently not being fertilised. "It is sometimes recommended to put pheasant hens into small enclosures open at the top, so that the wild cocks might get to them. I suppose generally that plan is successful, but in my own case it has failed entirely. I had plenty of eggs, but no chickens. My keeper gathered the eggs regularly and carefully, and they were duly set under common hens; but not one single egg came off. I know the wild cocks came close to the enclosure, but I never actually found one inside. I followed Baily's instructions implicitly; my own impression was, I must say, that the wild cocks had not visited the hens." This appears an exceptional case, and may probably be due to some local conditions.

On the other hand, a second authority states: "On an

estate with which I am well acquainted, the whole of the young birds, some 400, were reared from eggs produced by hens whose mates were wild birds. The pheasantry was constructed with an open top, and the wild cock birds regularly visited it. The tameness of these birds was remarkable, and I have frequently seen six or eight cock birds walking fearlessly about within a few yards of me while inspecting the birds. As an instance of the audacity of the wild bird, I may mention that a few years ago I kept five hen pheasants and one cock pheasant in a temporary covered pheasantry, the lower part being covered up to the height of two or three feet, and the upper part being constructed of wire stretched on poles. I noticed shortly after the birds had been put in that the wire was bulged inwards in several places, and could not imagine how it had been done. On watching, however, I found a wild cock pheasant was in the habit of regularly fighting with the confined male bird by flying up against the wire, the bird inside being by no means loth to accept the challenge. One morning, however, the wild bird was found inside, a nail having given way in one of his flights against the wire netting, being the cause of his unexpected capture. When discovered he had nearly killed the imprisoned cock bird, who was removed, and his adversary substituted. I may remark that those who have tried breeding from wild cocks will hardly, I fancy, return to the old system of keeping the cocks in confinement, as I have found that the birds bred from wild cocks are invariably stronger, and consequently easier to rear than those bred in the ordinary way."

There is no absolute necessity, however, for having recourse to the use of open pens, as the eggs of cut-winged birds, kept in pheasantries of sufficient size, well fed, with a good variety of fresh vegetable food, and supplied daily with fresh clear water, usually hatch quite as well as eggs gathered out of nests in the open covert.

The construction of more ornamental and permanent

aviaries has now to be spoken of, but will not require much consideration. Fixed aviaries are far inferior, as regards the health of the birds, to those that are movable, therefore, if possible, they should always be constructed so as to admit of their being shifted on to new ground as often as is convenient. The great cause of the comparatively small success that attends the rearing of pheasants in our Zoological collections arises from the fact that the birds are kept on the same spot year after year, and in aviaries that are not one-fourth of the size required for the health and comfort of the birds. The plan of an ornamental aviary necessarily depends on the desires of the owner, and hardly comes within the scope of this work. Mr. Crook, who had much experience in erecting ornamental aviaries, writes as follows respecting their construction: "A neatly constructed lean-to building may be employed, facing south or south-west; ten feet wide or long, six feet deep from back to front, and six feet high at front of the highest part of the roof; the roof should project over the side eighteen inches to throw off the wet. The ground must be dug out under the house, and dry earth or sand be filled in. Faggots may be placed here as before directed, or slanting against the back wall; every precaution being taken to induce seclusion for the nests. For those pheasantries desired for strictly ornamental purposes the run may be made to any size agreeable to the wishes of the owner and the conveniences of the ground at command; or of any design in character with some buildings near at hand. These ornamental aviaries may be carried out to any extent, but cannot be made to move about; therefore the greatest attention must be paid to any minute detail in construction to ensure the health and contentedness of the inmates. When it is possible, the pens or runs should be placed where there are some low-growing shrubs, or even currant or gooseberry bushes, as they afford good sheltering places, and it is quite possible that the hens will make their laying nests at the roots of some of them, which will be a benefit to the birds."

When the birds are left full winged in wire aviaries, and are wild, it will be found very advantageous to have a cord netting stretched some inches below the wire top, as otherwise the birds are very apt to injure themselves severely when they dash upwards on being alarmed. When it is required to handle the pheasants, precautions must be employed that are not needful in the case of fowls, for their extreme timidity causes them to struggle so wildly as often to denude themselves of a great portion of their plumage, or even to break or dislocate their limbs. They are best caught by the aid of a large landing-net, with which they can be secured when driven into an angle, formed by setting a large hurdle against the side or in the corner of the pen. Mr. Baily, in his practical little treatise, writes:—" The best way of catching them is with a net made of hazel rod, seven or eight feet long, forked at top. This fork is bent round, or rather oval shaped, forming a hoop long enough to take in the bird without injuring its plumage. It is then covered with netting loose enough to allow of its being placed on the bird without pressing it down to injure it, and tight enough to prevent it from turning round in the net to the detriment of its plumage. Where many birds have to be caught, it is expedited by the adoption of an expedient I will describe; and the plan is good, because it is always bad for the birds to be driven about, which they must be before they can be caught, if they are in a large pen. An extra hurdle should be made, to which a door should be joined on hinges. It should be three feet long. This should be placed by the side of one of those forming the pen, and the door being open the birds should be gently driven into it; then the door should be closed. They may then be taken with the hand or not. A pheasant should be caught with one hand, taking at the same time a wing and thigh, the other hand should be brought into play directly to prevent its struggling, and it may then be easily and safely held in one, taking both thighs and the tips of both wings in the hand at the same time. It takes two persons to cut the wings. They should always be held with their heads towards the person holding them."

Since the first publication of this work the plans advocated in it have been very generally tested and discussed. The remarks of one of the writers contain so many useful details that I am glad to reproduce the more practical portion of his letters.

"The advice offered with reference to pheasant pens or aviaries is as easy and inexpensive of adoption as it is good. By carefully following the excellent instructions fully set forth in the work upon pheasants by Mr. Tegetmeier—to whom the thanks of all lovers of the bird are due—I succeeded during the spring of 1875 in securing from thirty-five hens one thousand eggs. Forty birds similarly treated produced the following season 1500; last year forty-one hens presented us 1600; while this—so far as it has yet passed—offers promise of a still better return.

"The fertility of our eggs is most satisfactory, very nearly all proving fruitful, the few failing to hatch containing chicks, which through accident merely had not reached maturity. Here again I must gratefully acknowledge the excellent practical instructions proffered by Mr. Tegetmeier relating to feeding specially and management generally. We take all the pheasants with which our pens are supplied from early hatchings, care being observed that a due admixture of wild birds eggs are placed in these first sittings, thus securing a thorough change of blood.

"On or about Sept. 1 the young birds are caught up, the strongest selected, one cock to five hens, and, with a wing cut, placed in their future home. They require no further attention beyond the frequent supplying of fresh food and water twice or thrice a day, reclipping the cut wing excepted.

"Our aviary here being within easy flight of natural coverts, we adopt clipping in preference to pinioning, since, when the egg harvest closes, by extracting the crippled feathers, a gradual recovery of power enables the birds one by one to effect escape; the exodus thus permitted being generally fully accomplished in sufficient time for a thorough cleaning and preparation of the aviary in readiness for its proposed future young occupants. One of the great secrets of success lies in variety of dry and liberality of green food, together with a generous supply of frequently changed water, gravel or road grit, ashes, chalk, and pounded bones.

"I now propose offering a few suggestions touching more particularly the position, construction, and general management of the pheasant pens or aviaries. It may, however, be premised that their size and the numbers of birds proposed to be kept, greatly modifies many minor matters of detail, with reference not only to the health, but also to the comfort of the prisoners. On the all-important question of site —fair contiguity to the keeper's cottage should be observed; for if placed at too great a distance, a laxity, in winter more especially, of that solicitude so essential to their welfare, is likely to be engendered; while on the other hand close proximity, above all should there be many children, may, with all their custodian's care, prove the cause of great and irrevocable mischief. Total isolation, again, in the recesses of a deep, secluded covert, renders the birds so nervously sensitive that they are apt, upon the slightest unexpected excitement, to lose all self-control, dash about, and thus risk eggs, limbs, and even life.

"Our pens are placed within five yards of, and parallel to, a leading carriage drive, a thoroughfare daily in use. From earliest youth, therefore, the birds are more or less inured to the ever-changing sights and sounds incidental to ordinary traffic. Their thus seeing and hearing all going on around gradually enables them to acquire such an amount of courage, that curiosity usurps the place of fright; the cocks crowing joyously yet defiantly, while the hens peer inquisitively, yet fearlessly, through the lattice of their harems.

The pens should be sufficiently shielded by trees, so as to insure in very sunny weather a grateful shade; nevertheless, too much leafy shelter is apt to prove provocative of damp and cold. They should also, while enjoying a southern aspect, be well protected from the east wind. Thus placed the birds are better left without any well meant but fanciful attempts at further increasing their comfort. The little matters above enumerated excepted, the more they are exposed to the elements and permitted to rough it, the healthier and more robust will they become.

"As in our present case here, so it frequently occurs that insufficient space militates against that annual shifting of aviaries on to new ground, so often recommended, and upon which, so far as my experience serves me, where the utmost attention to scrupulous cleanliness has been observed, unnecessary stress is laid.

"After the laying season, when our birds have availed themselves of the liberty accorded them, the pens are completely denuded of their contents. The ground is trenched spade deep, thickly sown with unslacked lime, then covered with from two to three inches of fresh clean dry loam, and finally freely moistened with water through an ordinary garden-rosed watering-pot, when any floating lime dust is effectually disposed of, and the young birds may with safety be introduced.

"Our aviary, in its entirety, measures in width about 27ft., and length 108ft., there being, however, three transverse divisions, four square compartments are thus formed. A small trench, one foot in depth, is dug around the whole structure. A piece of stout wire netting, one foot six inches in width, placed with one edge in the bottom of the trench, has its other laced with wire to the hurdles, up the outside of which it extends nine inches, when the earth is filled in, and rammed. The inclosure is thus rendered fox, cat, and rabbit-proof; it has further attached to it 'gorse bavins,' thus securing warmth and privacy. The whole of the other

MANAGEMENT OF PENS.

portions have now strained over them stout 1½in. mesh galvanised wire netting, the top only carefully left free, for ingress and egress of wild birds. Inside each compartment, and parallel with the divisions, is now placed a row of bush bavins, one against the other, tightly pressed together, forming an inverted letter V. On the apex of these faggots the birds love to perch, preen, and doze, while a secure retreat in case of sudden fright is offered by the little tunnel left at the base. A few faggots may also for a similar purpose be placed leaning against the sides and corners of the inclosure, those angles where the doors are hung excepted.

"We have also two smaller pens, alike in all respects, and attached to those already described, but in measurement only 10ft. by 7ft. These are used for the temporary confinement of any quarrelsome egg-destroying or otherwise refractory bird, who can thus, until its wing is sufficiently strong for flight, remain. One of the hurdles dividing these small pens from their neighbours—as, indeed, in each of the interior divisions—should be easily removable to the end, that the birds can at pleasure be driven right through into the smaller pens for the purpose of capture, wing-clipping, &c.

"The introduction and placing about occasionally of freshly-cut fir tree branches is judicious. With reference to aliment, the greater the variety offered the better; and for a thoroughly trustworthy detail upon this vital point, again I gratefully add, *vide* 'Tegetmeier.' Regularity in the hours of feeding, however, is as essential as is the quality of food administered—three times diurnally, any unfinished *débris* of the previous meal having first been carefully removed, should the repasts be neatly and delicately served, not forgetting that, while all required is offered with no niggard hand, over-lavish generosity, only too often the mere promptings of laziness, ought most carefully to be avoided.

"Powerless are the prisoners to escape those fatal miasmatic vapours speedily generated by decaying vegetable and animal matter, which, when permitted to daily be trampled into the floors of the dwelling, are ever within a few inches, be it recollected, of their respiratory organs. In connection with this matter also, it is wise to have duplicate shallow circular galvanised iron water pans of about eighteen inches in diameter. They are light, and consequently more likely to undergo that thorough and frequent cleansing so necessary."

Coverts may be stocked either with wild birds or with those hatched in pens that have never been at liberty. Wild birds caught at the commencement of the year, not later than the middle of January, are healthier and more prolific than young birds that have never been allowed to fly. When caught, they should at once be put into large pens on fresh ground, having had the flight feathers of one wing cut off, when, if they are properly fed, they will become fairly tame before the

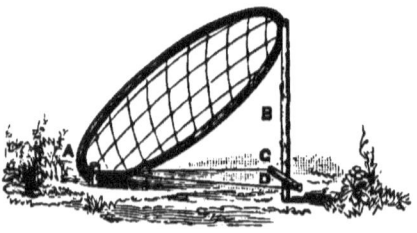

breeding season. However tame they may become they should not be kept more than one, or at the most two seasons, when their wings should be allowed to grow and other birds captured to supply their place. Other modes are adopted for capturing the wild birds. The above very simple form of trap is described by Mr. J. E. Harting, which is perfectly efficacious for the purpose required. It is merely a modification of the old-fashioned sieve trap, so arranged as to be self-acting, or, in other words, to require no watching. The accompanying sketch will make all clear. A is a hoop off a large cask, covered with slack netting. At the point where it touches the ground a peg is driven in, to which the hoop is tied, or, as it were, hinged. Another short peg is driven in at D, on the top of which rests a cross-piece C, above which again comes the long upright which supports the hoop. From each end of the cross-piece C, a piece of twine is carried to D, the twine being only a very little way off the ground. This acts as a trigger, and the moment a bird feeding under the hoop comes in contact with the twine, the cross-piece C is jerked away, and the trap falls.

Some breeders prefer large baskets six feet square by one foot deep, made of strong willow covered with canvas, to the sieve. This is propped up securely, and the pheasants feed under it for several days before they are caught. It is then raised by a single stick, from which a long wire or cord proceeds to a tree or shelter many yards distant. This is for the purpose of pulling away the stick and catching the birds that are feeding underneath it. Open crates are sometimes

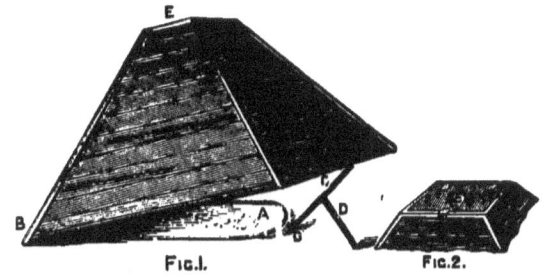

Fig.1.   Fig.2.

recommended to be used in the same way, but they are not advantageous, as the birds injure themselves in the endeavour to escape.

Another plan of a somewhat similar character, which has proved most successful in use, is the catcher represented in the above figure. It is made of deal, to be as light as possible, and can be painted brown. The size at the bottom should be about 2 ft. 4in. square, and at the top about 1ft. square, covered with a lid (Fig. 2), to enable the bird to be removed. To set it, as shown in the sketch, a bender is placed round from A to B, care being taken that it does not quite reach the front. Two sticks, C and D, are used, a notch should be cut in C about 6in. from the bottom, to admit the top of stick D; the lower end of C resting against the bender; and when the catcher is placed on the top of stick C the whole is held up by D, the bender being about 3in. from the ground. When the bird steps on the bender the trap falls and secures it. If the size described is used, the birds will hardly ever damage themselves. Where pheasants are to be caught, the catcher may be placed on the ground some time before using, propped up with one stick only, and some white peas strewn underneath, and nowhere else. With this trap it is no trouble to catch nearly every bird in the covert, however mild the season.

The best baskets for the transport of pheasants for short distances are those made of close brown wicker; in shape they should resemble a basin turned upside down, the part corresponding to the foot of the basin being uppermost, and forming the only opening into the basket. Before being used this opening should be covered with canvas, which is to be closely stitched down half way round, previously to the birds being placed inside, and firmly secured afterwards. In these baskets they are free from observation and molestation when travelling by rail or carrier, and from the baskets being close and circular they are much less liable to injure their plumage than when sent in more open and angular packages. In forwarding live birds care should always be taken to attach a stout and somewhat loose cord across the top of the basket, in order to serve as a convenient handle by which it can be lifted with one hand, otherwise, in the hurry of transit, the railway porters, who cannot be expected to use both hands in lifting every package, are certain to catch it up suddenly by one side, and the birds are often severely injured by being suddenly and violently thrown against the opposite one.

The consideration of the best means of arranging for the transport of birds over long distances and by shipboard, is given at length in the Appendix.

CHAPTER VII.

MANAGEMENT OF PHEASANTS IN CONFINEMENT (CONTINUED).

LAYING AND HATCHING.

F the laying in aviaries there is but little to be said. The birds usually drop their eggs about at random, consequently they should be looked after, and collected frequently, so as to prevent as far as possible their being broken, which is almost certain to establish the destructive habit of egg eating. Sometimes, however, hen pheasants will take to concealed nests, and instances are not unknown of their sitting and hatching successfully in confinement. A correspondent states: "In 1852 I had a cock and three hens in a small place (I will not dignify it by the name of an aviary, for it is open at the top, and the birds are pinioned or have their wings cut); one of the hens made a nest, and sat and hatched five young ones. These, unfortunately, the other pheasants killed directly they came from under the mother. In 1853, the same hen sat again on eleven eggs, and hatched seven, when I let her out into my small garden, and a better mother I never saw; she would allow no strangers to come near her without flying at them. At the end of seven weeks the gapes killed them all. It was a curious sight to see the old pheasant make her nest of ivy-leaves and hay, the former of which she always used to cover her eggs with when she left her nest, doing so by standing on the edge, and throwing the leaves over her back. The same hen sat again in 1854."

Mr. G. F. Woodrow (Keeper to the Earl of Denbigh, Newnham Paddox, Lutterworth), writing on the subject stated: "I have half an acre of young plantation inclosed for a pheasantry and open at the top, so that the wild cock birds can go in and out. I had over thirty hen pheasants and three cocks, all with their wings cut. About ten weeks ago a hen pheasant wanted to sit on the last egg that she laid; I took it from her, and disturbed her every day, but she persisted in sitting without an egg for more than a week; at last I took pity on her. One evening when I had gathered the eggs I put sixteen under her, and she sat and hatched thirteen birds. She allowed me to lift her off the nest, and I took her and her young and put them in a hen coop, and she has reared them well, and, quite as tame as any of my hens that I have rearing pheasants, allows me to drag the coop on to fresh ground, and never flutters. As soon as I throw the food in front of the coop she commences calling her young. They are now about the size of landrails, and the whole of them living."

To prevent the fatal habit of eating the eggs, no care should be spared, as it is entirely subversive of any hope of success in rearing. As before stated, it may be in great part prevented by the frequent collection of the eggs. Mr. F. Crook truly remarks: "The male bird in confinement frequently takes to pecking the eggs, at first only for want of something more natural to do. Having no space, no fields and copses to roam about and amuse himself in, he pecks and pushes the egg about. At last it gets chipped, and he tastes of its contents, and he will not then leave it until consumed, and the abominable habit is confirmed in him. As it is usually the male bird that commits these vexing faults, a loose hurdle forming a corner pen, into which he can be driven, will be found most useful, as he should only be allowed amongst the hens after they have laid their eggs for the day; and all having been removed, a wooden egg may be exchanged for the real one, which will soon tire him out; and the bad habit may be cured, and no loss of time occur in the breeding season. But whether the birds are troublesome or not in this respect, the attendants must make periodical visits to the breeding pens for the purpose of collecting the eggs, as they should never be allowed to remain about."

There is no doubt that bad management and improper feeding tend to promote this serious evil. The frequent disturbance of the birds by the inquisitiveness of visitors, bad and improper stimulating food, without a sufficiency of green vegetable diet, want of cleanliness in the pen, an insufficient or dirty supply of water, and want of grit to assist digestion, all aid in developing the habit. Mr. J. F. Dougall, in his "Shooting Simplified," suggests the following mode of preventing the practice when once established: "In pheasantries means should be taken to prevent the eggs being destroyed by the male bird; and as it is impossible to keep continual watch, the hen should be induced to seek a dark secluded corner by forming for her an artificial nest covered thinly with straw. Under this straw have a net of mesh exactly wide enough to allow the egg to drop through into a box below, filled with soft seeds or shellings, leaving only a few inches between; the cock bird cannot then reach the egg, which falls uninjured on the soft seeds below, and is safely removed."

Mr. Leno writes: "I have invariably found the cocks to be the culprits. As soon as a pecked egg is found, the cock bird should be removed, and the hens left by themselves for a few days, to see whether he is or is not the guilty one; before putting in another cock with the hens, fill up the shell of the broken egg with soft soap, which the fresh bird may try his hand at. In case the first cock has been at mischief long enough to teach the hens, there is no saving the eggs, unless they are watched and the eggs picked up immediately they are laid, or by partitioning part of the pen off, and straining some galvanised wire netting across the inclosure six inches off the ground, the mesh being of a sufficient size to allow the eggs to drop through as soon as laid on to some moss or chaff; the hens should be driven into the wired inclosure early in the morning, and let out again late in the evening—food and water, of course, must be placed in a small trough for them."

Mr. Fairfax Muckley, of Audnam, Stourbridge, says: " My pheasantries are large, and of considerable extent. My method is this: In the beginning of April I have a bundle of larch bushes placed on each corner of the pheasantries, leaving only room behind for one bird, and a little hole in the bushes for the hens to creep into; then make a place on the ground behind the bushes and put two or three sham ground glass eggs, and also place a few anywhere about the pheasantries; they then become accustomed to see these sham eggs and try to break them, but finding they cannot do so, they leave the real ones alone. The hens are also induced to go into the corners of the pheasantries and lay to the sham eggs. The great thing is to have these in every way like real ones. Those generally used are useless, being either too heavy or too light, and wrong in appearance. I may add that the oftener the eggs are collected the better; but care should be taken not to disturb the hens when behind the bushes. I had two very fine cock birds sent me; they ate the eggs in the beginning, but by continually having perfectly-made sham eggs before them they are quite cured, and over one hundred eggs have been collected out of their pens. It is a good plan, when a hen has just laid, to take the egg away and put a sham one in the place, particularly when you know they eat them. At the end of the season have the sham eggs collected for other seasons."

The glass eggs manufactured by Mr. Muckley are most efficacious in preventing this destructive habit.

In consequence of the removal of the eggs as soon as deposited, and the birds not sitting, the number laid by the hens in confinement is greatly in excess of that produced by them in a wild state, sometimes as many as twenty-five or thirty being laid by one hen. This extreme prolificacy tends to exhaust the birds, and it will be found most advantageous to turn them out when they have finished laying, and to supply their places by young poults.

It not unfrequently happens that a greater number of eggs are required for hatching under farmyard hens than are produced by the birds in the pheasantries; in such cases the surplus eggs in the nests of the wild birds may be advantageously collected. This, however, may be done in a right or a wrong way. They should be taken before the hen pheasant begins to sit; and if removed one at a time every other day as the bird is laying, they are certain not to have been partly hatched.

Richard Jeffries, in a most graphic article on the pleasures of pheasant rearing, describing the gathering of the eggs, truly says: " Unfortunately nothing is more easy to find than a pheasant's nest. Like a cockney looking for a home in the suburbs, the hen pheasant seems to prefer a lively situation near a thoroughfare, with a good view of anything that may be going on. It needs no great practice to catch the glance of the bright beady eye among the roots of the roadside hedgerow, or to distinguish the grey mottled plumage among the grass and nettles in the ditch below. Look under that heap of fallen boughs, and as likely as not there are the green-grey eggs dropped under the very outermost, where there is scarcely a pretence at cover, although, had she taken the trouble to force her way one half-yard further, the hen might have laid them safe out of sight of all but ground vermin. So by dint of poking about among the grass and the branches and brambles, by looking under furze bushes and in hedgerows, and in the cavities formed at the foot of tree trunks, you may come upon a good number of nests in the afternoon, should birds be tolerably plentiful. Very likely indeed you have found too many eggs to be accommodated under the sitting hens at your disposal. Some must be left, while other brood mothers are sought. Whether on your second visit you find those you left, as you left them, depends greatly upon circumstances. If you have a profusion of rooks about your place, the chances are much against it. For those omnivorous gluttons have as decided a partiality for pheasant eggs as any ball-going gourmand for those of the plover. They have overrun your woods. They sit swinging and cawing on each projecting bough that commands a prospect. They walk the slopes of your fields, one eye closely scanning the soil for insects, the other sweeping all the points of the compass. Nothing escapes their observation. When they see you out for an object they follow you and mark each movement. We have very little doubt they speedily learn to suspect your intention, and when they see you stoop in a likely spot, they fly down to institute an investigation whenever your back is turned. In no other way can we possibly account for the wholesale wreck of eggs that had been spared and sat upon until you visited them in your walk. And if you doubt who are the culprits, try the ordeal by taste, and strychnine a nestful of eggs. You will find the bodies of the black delinquents strewed round the fragments of the shells.

"Nothing can be prettier than the broods of young pheasants as they are hatched off, tame as chickens—although more graceful and active—running from the shell, and beginning forthwith to peck about for a living. Unfortunately there are other members of the animated creation who watch their growth and their movements with even keener and more immediate interest than yourself. For some four months to come you mean neither to shoot nor eat your confiding protégés; but they are surrounded by sharp-set carnivora who propose themselves that pleasure on the earliest possible opportunity. We do not assert that those nuisances the rooks are dangerous in this stage of the pheasant breeding, although we should deem it imprudent to trust them too far. And there a weasel is watching, popping his head at intervals out of different holes in the neighbouring bank, undeterred by the fate of several of his family, who have already been trapped there and gibbeted. But more dangerous than hawk or weasel are the jackdaws. For, as these vociferous birds bear comparatively respectable characters, they are more likely to be indulged with a licence they abuse. We know them to be *bavards :* we cannot deny the family tendency to kleptomania. But we are in the way of believing chattering to be the sign of a frank, shallow nature, and we are apt to condone the thefts that are perpetrated with no view to profit. In reality, the jackdaw is a deep hypocrite—a robber and a bloody-beaked murderer. He chatters his way from branch to branch above the coops with the most unconcerned air in the world—just as a human thief walks, whistling, with his hands in his pockets, towards the prey he means to make a snatch at. Then, when he sees himself unnoticed, the jackdaw stills his chatter and makes his stealthy swoop; and in this way, watching while your watcher's back is turned, he massacres a whole family of your innocents, and the hawks and weasels get the credit of the crime. But, after all, a gun kept upon the spot generally inspires a salutary dread.

"Many of your young birds survive the perils of their cheeperhood; then the long grass in the neighbouring bits of covert becomes alive with them, and once in that stage they are comparatively safe. Thenceforward till the autumn they feed and thrive, strengthen and fatten. And, sport, sale, and the autumn game course out of the question, what can be pleasanter or prettier in the way of sounds or sights than the young birds learning to crow in your coverts as you saunter out before breakfast, or scattered about your lawn as you dine, with open windows, of a summer evening ? *Pace* Mr. Tegetmeier, and other gallinaceous authorities, we must say that in the way of pets we prefer pheasants to poultry."

Many pheasant rearers are so short-sighted as to recruit their stock of eggs by purchase, forgetting that in the great majority of cases these eggs are stolen, either from their own or from other preserves. In some cases the keepers themselves purloin the eggs and sell them to the dealers, from whom they are perhaps repurchased by the owner of the very estate from whence they were abstracted. As an example of the mode in which these frauds are perpetrated, I may adduce the following example, furnished by a correspondent: "On a small estate in Sussex there was a pheasantry with about seventy-five birds, and when the laying commenced, the eggs were taken up carefully two or three times a day; the keeper had these eggs out as he got the hens ready to sit, which was three or four times a week, as a very large number of hens were kept. A book was kept, in which were entered the eggs laid each day, the eggs given out being also entered in a second column, and the number of birds hatched in a third; and the keeper was directed to preserve all the eggs not hatched or bad, so that they might be added to the number of birds, and the total of birds hatched and bad eggs compared with the eggs laid. The first ten or twelve hens brought out good broods of from thirteen to seventeen birds each. Afterwards they decreased, and in many cases there were only three, and even as low as one bird in a brood. The eggs were never more than a day or two old when first sat upon, we had often hens waiting for the eggs, and everything was most favourable for a large return of birds. At this time some suspicion was entertained, and for a time the keeper was more closely looked after, when the broods at once came up to twelve and fourteen birds. But, unfortunately, the same watchful care was not continued, and at the end of the season it was found that he was short upwards of seven hundred eggs, and that he had sold upwards of thirty-five pounds worth. The sitting-house was a first-rate one for the purpose—large, roomy, and dry. The keeper's plan was to keep back a portion of good eggs out of each setting, and substitute bad ones in their place. I am very far indeed from saying that this is a

SELECTION OF HENS FOR HATCHING. 103

common occurrence; for I am glad to say that most keepers are as anxious about their charges as their employers, and take a pride in showing a large head of game."

From the indisposition shown by the pheasant to incubate in confinement, it is necessary in all cases to have recourse to the hens of the domestic fowl as foster parents. Various opinions are offered as to the breed of fowls most suitable for the purpose. There can, however, be no doubt that it should be one of a moderate size, and not too prolific in egg producing, as it is essential that the mother hen should keep with the poults as long as possible, which she is not likely to do after she commences laying. Silky fowls are strongly recommended by some, and they unquestionably constitute admirable mothers. M. Vekemans, of the Antwerp Zoological Gardens, where rare pheasants are reared more successfully than in any similar establishment in Europe, employs half-bred silkies; and the late Mr. Stone, of Scyborwen, fully indorsed his opinion. These half-bred silkies are good sitters, admirable mothers, and keep a long time with the young. The ordinary bantams sometimes recommended are undoubtedly too small, not being able to cover the poults when of any size. The employment of pure bred game hens is strongly recommended by many breeders of pheasants, as they will defend their chicks against any enemies that may attack them, though their natural wildness renders their management somewhat difficult at times; any small, tame, ordinary hens will answer if known as good nurses, and none others should be employed.

Hens with feathered legs are not desirable, as they are very frequently afflicted with what is known as "scurfy legs," a very obnoxious disease, which is caused by minute parasites that breed under the scales, causing rough swellings. These parasites extend to the young pheasants, and many coverts are infested with scurfy-legged pheasants in consequence.

It is the common custom to set the hens in close boxes, with little or no ventilation, crowded together in sitting houses. Under these conditions the nests swarm with vermin, the sitting hens become irritable and break their eggs; and when the young pheasants come out they are infested with fleas and lice, and are nearly devoured alive. Moreover, the dry, stifling air of these places is destructive to the vitality of the unhatched birds, numbers of which die in the shell either before or at the period of hatching. Every poultry keeper knows that no nests are so prolific of strong healthy chickens as those that the hens "steal" under hedges or in copses or concealed places, from whence they emerge with strong flourishing broods that put to shame the delicate, sickly youngsters reared in the close air and dry over-heated nests of a hatching-house. The nearer we can imitate Nature the better—and if the hens hatching pheasants' eggs can be set on the ground, covered over with a ventilated

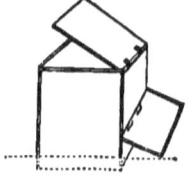

HATCHING BOX.

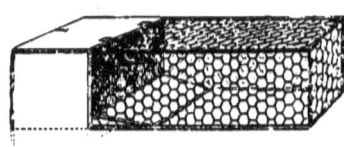

BOX AND RUN COMPLETE.

coop—more for concealment than warmth—and this surrounded by a wire run, into which the hen can come out, feed, drink, and, above all, dust herself, at her will, the eggs will be found to hatch out much more abundantly than when they are set in the vermin-infested, crowded pigeon holes adopted by many keepers. Such nesting boxes may be a cheaply constructed box, as shown in the woodcut. The nest should be on the ground, there being no bottom to the box; and if the sides and the wire work are sunk into the earth, and the latter is sparrow and rat proof, the hen may be supplied once daily with food and water without entailing any further trouble. But some dry ashes should be given in which she can dust herself, and it is needless to say that the larger the wire inclosure can be made the better.

In confirmation of my views on the subject of hatching, I have much pleasure in quoting the following practical observations of Mr. F. Crook, who states :—" The fault usually existing is, that an over-careful, pampering system is adopted, and miserable broods are the result. I have experimented in a manner which leaves no doubt upon the subject. Upon one occasion I was anxious to test the fertility of certain pheasants' eggs, and continued to remove the eggs from a nest in the woods until I found the hen desirous of sitting. I left twelve eggs in the nest, and I sat thirteen at home under a hen; the pheasant brought out twelve birds, while at home I only had three miserable birds. Similar results have many times occurred since. As a rule, the home hatching places are too confined in area, the hens are fed too near the nests, and are not compelled to remain off the eggs long enough, and no amount of wetting or sprinkling with water, either hot or cold, recommended by some writers, will compensate for a due supply of fresh air. Birds in the woods select a dry spot, sheltered from the rains as much as possible. Sometimes they will carry dry leaves, soft, short straw, hay, and feathers; at other times the nest is made in a hollow at the root of a tree, and the eggs are laid on the loose mould; or under thick bushes, and covered with coarse grass; but in every case the nest is *never stifled,* having the freest circulation of air surrounding it. If such natural precautions alone are used, greater success may be looked for at home than when the nests are made up in quiet, warm, small places, where the birds have but little room to move, and the eggs get nothing but a fœtid atmosphere to destroy the life that lies beneath the shell. The term of incubation of pheasants' eggs varies considerably. I have hatched them at home at all times from twenty-two to twenty-seven days, but in the woods they invariably turn out about the twenty-fourth day. Those which hatch at the most natural time of twenty-four days turn out to be the finest and healthiest birds. There is some care required in marking the dates and number of eggs set in each nest for hatching, as by a little forethought in this respect, great advantages may be obtained by saving time, and retaining the services of the sitting hen. Over each nest the date should be distinctly pencilled, thus $\frac{14}{16-4-96}$ which means fourteen eggs were set on the 16th of April, 1896. About the ninth day the eggs should be examined, and all those which appear perfectly clear, as when first set, should be laid on one side as useless for hatching, but as perfectly good for feeding the poults."

This examination of the eggs after they have been sat on for a few days is exceedingly desirable, as those that are unfertilised may be removed, when they serve as food for the poults, and leave more room for such as contain live birds.

Many instruments dignified by the title of ovascopes and egg-testers have been devised for this purpose, some with lenses, others with reflectors, &c. I have tried the whole of them, and find them far inferior to the following simple contrivance, the description and engraving of which is reproduced from my work on "Table and Market Poultry":

"The most simple egg-tester is made out of a piece of cardboard; the cover of an old book answers very well. An oval hole should be cut in it, not quite large enough to allow an egg to pass through, and if the cardboard is white, one side should be inked or painted black. The eggs are more conveniently removed from the hen at night, or if in day they should be taken into a room from which daylight is excluded. A single lamp only should be used. The cardboard, with the darkened side towards the observer, should be held near the chimney of the lamp, and the eggs one after another, should be held against the hole. Those that contain chickens will be observed to be quite dark and opaque, except at the larger end, where the air-space exists. These should be replaced under the hen. Those that have not been fertilised, and are consequently sterile, are sufficiently transparent to allow the light to pass through, and look as fresh eggs would if examined in the same manner. Such eggs are usually termed 'clear.' These clear eggs are perfectly good to eat; but it is preferable to save them for the food of the chickens when hatched. Throwing them away is a wasteful proceeding."

It is evident that setting two or more hens on the same day is advantageous, as the "clear" eggs may be removed from the whole of the nests, and the number in those that are deficient made up from the other nests, a fresh batch being

METHOD OF TESTING EGGS.

placed under the hen the whole of whose eggs have been removed.

The conveyance of eggs for the purpose of hatching is tolerably well understood by the most experienced breeders. There is nothing equal to a good-sized basket in which they can be placed, surrounded with and separated from one another by hay. Boxes with bran, sawdust, cut chaff, &c., are very inferior, as these materials shake into smaller compass by the jolting of the journey, and the eggs frequently come into contact and are broken.

Sometimes circumstances may occur in which it is desirable to exchange the eggs of fowls and pheasants temporarily; there is no difficulty in so doing. Pheasants' and partridges' eggs may be taken from their nests, and others substituted. The exchanged eggs may be placed under common hens. As soon as the pheasants' eggs show appearance of hatching, they are removed back again to those nests which have not been forsaken, with very good results. The exchange is much more likely to succeed with pheasants than partridges; with the former it is almost a certainty. The advantages are many, and all on the keeper's side, as he may turn out with the old birds larger broods than they otherwise would have hatched.

In those cases in which the nest of the pheasant is in a situation likely to be disturbed, the plan may be advantageous; but, under ordinary circumstances, the eggs had better be left unmolested, as the hen pheasant is almost certain to bring off a larger number of chicks than would result if the eggs were shifted under a farmyard hen.

In some parts of Germany turkey hens are employed to hatch pheasants; the eggs are collected and placed under the hens, who make excellent mothers, and are capable of hatching and rearing twice the number of poults that a barn-door hen can raise. From the great success that has attended the introduction into England of the American plan of allowing turkey hens to lay, sit, and rear their young in the open, I should strongly advise the placing of pheasants eggs in the nest of a turkey hen that has sat herself in some hedgerow or covert, and letting her rear the young pheasants, uncooped, and at perfect liberty.

CHAPTER VIII.

MANAGEMENT OF PHEASANTS IN CONFINEMENT (CONTINUED).

REARING THE YOUNG BIRDS.

SUCCESS in the rearing of young birds, it cannot be too strongly impressed on the inexperienced pheasant rearer, is never the reward of those who practise perpetual intermeddling with the sitting hens. All interference at the time the eggs are hatching is injurious; nevertheless, there are fussy people who cannot imagine that anything can progress rightly without their assistance; when the eggs are chipping they disturb the fowl to see how many are billed; this is generally resented by the hen, who sinks down on her eggs, and most probably crushes one or two of them, and thus renders the escape of the young birds almost impossible. It is perfectly true that sometimes an unhatched bird that would otherwise be unable to extricate itself, may be assisted out of the shell and survive, but it is no less certain that for one whose life is preserved in this manner a score are sacrificed to the meddling curiosity of the interferer.

The chicks should be left under the hen till they are twenty-four hours old without being disturbed; by this time the yolk which is absorbed into the intestines at the period of hatching will have been digested, and the young birds become strong enough to run from under the parent hen.

If the fowl is set in one of the coops with a wire run, such as I have recommended, she had better be left alone, and will leave the nest herself as soon as the chicks are strong enough to follow her. The ridiculous practice of taking the young birds as soon as hatched, dipping their bills in water or milk to teach them to drink, and forcing down their delicate throats whole pepper corns or grains of barley, is so opposed to common sense that it does not need to be refuted. When young pheasants and fowls are hatched in a state of nature, they are much stronger and more vigorous than those reared under the care of man (unless, indeed, the season be so wet as to be injurious to the wild birds), although they are not crammed with pepper corns, highly spiced artificial foods, and other nostrums, but have to seek their first food for themselves. Nature is far cleverer than man, but, unfortunately, the latter has not always the sense to perceive the fact. The nearer we can imitate her in our arrangements, the more successful we shall be.

With regard to the first food of the young chicks, there is nothing superior to a supply of fresh ants' eggs (as they are generally termed, although, strictly speaking, they are the pupæ, and not the eggs of the insects). For grain, I am anxious to recommend, as the first food, a good proportion of canary seed in preference to grits and meal. Grain when once crushed or bruised has its vitality destroyed, and it then undergoes changes when exposed to the air: the difference between sweet, new oatmeal and the pungent, biting, rancid meal that is often found in the fusty drawers of the cornchandler, is known to all persons accustomed to use oatmeal as food. This change, however, does not occur in the entire grain as long as its vitality exists, and hence the whole canary seed, which is readily devoured by the young pheasants, is almost certain to be fresh and sweet. Moreover, the husk contains a larger proportion of phosphate of lime, or bone-making material, than the centre of the grain, and is, therefore, better adapted to supply the wants of the growing birds. The first food preferred by young partridges is the seed of the crested dog's tail grass (*Cynosurus cristatus*), with which their crops will often be found quite full, and there is no doubt it would be an equally advantageous food for young pheasants, but is not as readily obtained as canary seed.

To afford a supply of artificially prepared animal food, most of the books recommend hard boiled eggs, grated or chopped small, to be mixed with bread crumbs, meal, vegetables, &c. Nothing, however, can be less attractive to the young birds than the food they are frequently condemned to exist upon. I have often seen pieces of the chopped white of hard boiled egg, dried by the sun into horny angular particles, refused by the young birds, although on these, with bread crumbs also dried to brittle fragments in the sun, many persons attempt to rear young pheasants—and fail. The best substitute for ants' eggs is custard, made by beating an egg with a tablespoonful of milk, and " setting " the whole by a gentle heat, either in the oven or by the side of the fire. The clear eggs that have been sat on for a week answer perfectly well. No artificially prepared animal food can surpass this mixture. The egg supplies albumen, oil, phosphorus, sulphur, &c.; whilst the milk affords caseine, sugar of milk, and the requisite phosphate of lime and other mineral ingredients; moreover, these are all prepared and mixed in Nature's laboratory for the express purpose of supporting the life and growth of young animals, and combined as custard form a most soft, sapid, attractive food, that is eagerly devoured by the poults. From my own long experience in rearing many species of gallinaceous birds, I am confident that a very much larger proportion can be reared if custard forms a considerable proportion of their food for the first few weeks, than on any other dietary whatever.

Many rearers of pheasants are strongly in favour of using curd, made from fresh, sweet milk put on the fire, and when warm turned or curdled with alum, and then put into a coarse cloth, which is to be twisted or pressed until the curd is a hard mass. There are several objections to curd as food. The alum is a powerful astringent, and is not a natural diet for young birds. The curd so made only contains two of the constituents of the milk, namely, the caseine and the cream. The whey, containing the sugar of milk, the saline ingredients, and, above all, the bone-making materials, is rejected, whereas, when the milk is made into custard, the whole of the constituents are retained, and to them is added the no less valuable ingredients of the egg. There is, in fact, no comparison to be made between the nutritive values of curd and custard.

Gentles or the maggots of the bluebottle or flesh fly are used by some keepers. They are generally obtained by hanging up in the woods, at a distance from a human habitation, some horseflesh, or the bodies of vermin that have been killed, and the gentles are allowed to drop into a tub of bran. The plan is necessarily offensive. A much better plan in situations where it can be employed, is to allow the dead bodies of any animals to become thoroughly fly-blown, and then to bury them a few inches in the soil, as described at page 83. It is obvious, however, that this plan cannot be pursued where the pheasants are reared under hens confined in coops. Maggots can also be procured in the neighbourhood of the sea coast by adopting the following plan, recommended in Cornwall Simeons' "Stray Notes on Fishing and Natural History:"

"It is not, I think, generally known that maggots admirably adapted for feeding young pheasants and partridges can be procured from common sea-weed. This should be taken up as near low water mark as possible, placed in a heap, and allowed to rot about a fortnight, after which it will be found swarming with maggots, rather smaller than those bred in flesh. The keeper from whom I learnt this dodge, a man of considerable experience in his vocation, tells me that he considers them, as food for young birds, superior to flesh maggots, inasmuch as they may be given in any quantity, without fear of causing surfeit."

When the hens are cooped, as is necessary where numbers of pheasants are reared, a good supply of fresh vegetable food is absolutely necessary; and I believe that nothing surpasses chopped lettuce, which should be running to seed, and consequently milky, as the pheasants take to it much more readily than they do to onions, watercress, &c., or other green food. The greater the variety of food the better; therefore, in addition to the articles before spoken of, a little crushed hempseed, millet, dari, and coarse Indian corn meal, if fresh, &c., may be added.

As the mode of treating pheasant chicks by different breeders varies considerably, it is desirable that I should indicate the management which has been found successful in other hands. I will first quote the practical directions of Mr. Bartlett, the superintendent of the gardens of the Zoological Society, Regent's Park. This paper was written for Mr. D. G. Elliot's "Monograph on the Phasianidæ," and I beg to return my thanks to these gentlemen for permission to quote it *in extenso*. Mr. Bartlett writes: "At first the chicks require rather soft food, but not very moist. One of the best things to give them is hard-boiled egg grated fine, and mixed with good sweet meal, a little bruised hempseed, and finely chopped green food, such as lettuce, cabbage, watercress, or mustard and cress. Meal mixed with boiled milk until it is like a tough dough, sufficiently dry to crumble easily, together with a small quantity of millet and canary seed, is also excellent for them. A baked custard pudding, made of well beaten eggs and milk, is likewise of great service to the young; and if the season is wet and cold, a little pepper, and sufficient dry meal to render it stiff enough to crumble, should be added before baking. Ants' eggs, meal worms, and grasshoppers are also very useful. The first of these are easily obtained in a dry state, in which condition they can be kept many months, and are invaluable.

Care should be taken that fresh and finely-chopped green food should be given daily. Many persons are in the habit of giving gentles to young birds; there is great danger in these; and I merely mention them, without recommending their use; for, unless the person who gives them will take the trouble to keep them for some time in moist sand or damp earth until they have become thoroughly cleansed, they are apt to cause purging. Many valuable birds have been lost by the incautious use of gentles freshly taken from the carcase of some dead animal; but, if well cleansed by keeping ten or twelve days after being removed from the flesh, a few, very few, may be given in case no better kind of insect food is at hand. The treatment of the young birds, such as change of food, &c., must greatly depend upon the judgment and skill of the person who has charge of them. Much also depends upon the locality, the state of the atmosphere, the temperature, the dryness or wetness of the season, the abundance or scarcity of insect food, and other considerations which must serve to guide those in whose care the chicks are placed."

The mode of management pursued by the late Mr. Douglas is somewhat different. He truly remarks: "Although food has a great deal to do in the rearing of pheasants, attention has almost an equal share; and without the attention required being given, food would be of little avail. I will commence with the hatching. Never remove your hens until the chicks are well nested, guarding the nest to keep any that may be hatched before the last chick is strong enough to leave the nest. Never take the first hatched from the hen—it is wrong; nothing is so beneficial in strengthening a chick as the heat of the hen's body. Let feeding alone for the first twenty-four hours after the first chick is hatched; the large quantity of yolk that is drawn into the chick within the last twenty-four hours of its confinement in the shell is sufficient for its wants during the time specified. Next, have your coops set on dry turf two or three days previous to your pheasants being hatched; it will save a little hurry when wanted; also it will keep the spot dry, that being so necessary on the first shift from the nest. If your turf is not of a sandy nature, sprinkle a handful of sand on where you intend to shift your coops. The coops being shifted daily is very beneficial to the chicks. Take care they are not let out in the morning until such time as the sun is well up, if there is a heavy dew on the grass, and the grass has got a little dry. I have no doubt but the continual letting out on wet grass, previous to the sun having power to counteract the bad effects of the cold wet dew, is the cause of many of the ills they are subject to. Feed twice or thrice, if necessary, previous to letting out. The principle food I give for the first fortnight is composed of eggs and new milk, made as follows: In proportions, one dozen of eggs, beaten up in a basin, added to half a pint of new milk; when the milk boils add the eggs, stirring over a slow fire for a short period to thicken, when it will form a nice thick custard. This I give for the first three days; then I commence to add a little of the best oatmeal, and any greens the garden can produce, finely chopped, for the next three or four days; after seven days I add to their diet a little kibbled wheat—being kiln-dried previous to kibbling—also split groats and bruised hempseed, occasionally a handful of millet seed; taking care all their food is of the very best, and that the feeding dishes are scalded in boiling water daily. The above food I use until about three weeks old, when I add minced meat mixed with oat or barley-meal, with the broth from the meat, the meat being composed of sheep's heads and plucks, taken from the bone and finely minced, and just sufficient of the broth to form a dry crumbly paste. At five weeks old I consider a feed of good wheat and barley alternately, the last thing at night, quite necessary, not forgetting, at this age, to add a little tonic solution of sulphate of iron to their water daily. At this time the growth of their feathers requires a great deal of support, and if the bodily strength is not supported by a strengthening diet, they must give way. Continue the custard up to eight weeks old, but adding more meal to it, with the green food. Give one sort of food at a time (just so much that they eat it clean up), and attendance every hour from the time you commence to feed until shut up for the night. Change the water repeatedly during the day."

With regard to the coops employed for the hens with young pheasants, a form much recommended is one made like a box, 3ft. long, 2ft. wide, and 2ft. high in front, sloping off to 1ft. high at the back, and having a movable boarded floor that may be employed if the ground be wet. The birds ought to have a further space of about two yards square to run in, fenced in by sparrow-proof wire netting. A good coop of this kind is shown in the cut. The inclosed run,

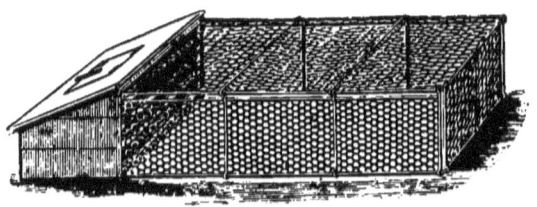

which is proof against rats and sparrows, &c., affords a sufficient space for the exercise of the young birds a day or two after hatching, after which the coops should be placed without the wire runs in the spot where the young birds are to be reared, the grass, if high, having been mown around some short time previously, so that the young shoots and tender clover may be growing for the use of the birds. Mr. Reynolds, of Old Compton-street, has some admirable coops of a similar kind. The advantages of these arrangements have been very ably set forth by Mr. T. C. Cade, of Spondon, Derby. He writes: "There is a great saving of food, as small birds are excluded by the wire netting; and it is also practicable to put down a good supply of food at night, so that the young pheasants may be able to feed as soon as they wake, and not be kept waiting, according to the usual plan, for two or three hours during the long summer mornings before they are let out. My birds are never shut in the coop at night, the wire netting being sufficient protection against vermin and cats. I do not know whether any of your readers have ever accompanied their keeper on a hot summer morning when he is letting the young birds out of the coops. If not, let them do so, and but put their noses within a foot of the coop, and I will venture to say that they will never allow such cruelty again. More than a dozen birds confined, perhaps, for ten hours in a dirty, ill-ventilated box, containing less than half a cubic yard of air. No wonder that they look languid and drooping, and that it takes them half the day to recover. I am far from insisting that the birds should at all times be kept in these small yards. When they are more than a week old I would, in fine weather, raise one of the sides and let them roam at their will, of course, replacing the board at night. But in wet weather and in the mornings before the dew is gone, I would keep them up, and not run the risk of their getting draggled and chilled with running on the wet grass." When shut in at night, which is often necessary to avoid loss by weasels or rats, &c., they should be let out at daybreak in the morning.

Many keepers prefer rearing the young pheasants under hens that are tethered by a cord to a peg driven into the ground, with an open shelter coop into which they can retreat, and if necessary be shut in, at night and during rain. The following directions for tethering hens with young pheasants or chickens are taken from my work on "Table and Market Poultry."

"The hen should be fastened by a piece of string to a peg driven into the ground, and an open, sheltered coop should be placed near her, under which she can retreat at night and during rain. The coop should not be put so close to the peg to which the hen is attached that she can walk round it, but near the limit of her cord, so that she can pass in and out, but not round the back. Under these circumstances she will be able to scratch the surface of the ground and supply her young with the seeds, grubs, worms, and natural food, which is so much more advantageous to them than any artificial substitute that can be given. The little chickens, even when two or three days old, will be observed scratching for themselves, and the progress that they make when reared under these conditions is out of all proportion to that made when the hen is kept cooped up, and the birds are fed on the hard, soiled, dirty ground.

"But there is a right way and a wrong way of doing everything. If the hen is simply secured by a piece of string tied round her leg, she will pull against it, and the leg or upper part of the foot may be injured. She should be secured by a proper jess, such as is used by falconers. A piece of thin, flexible leather, about eight inches long, by something less

than 1in. broad, should be taken, and three openings cut in it, as shown in the diagram, which is one-half the required size. The part between A and B should be placed round the leg of the hen, the slit A being brought over B, then the end C should be passed through both slits, care being taken that it goes through A first. It should be pulled right through, when it will be found to make a secure loop round the leg of the hen, which she can neither undo by picking nor tighten by pulling. The cord of the requisite length is then tied to C, and fastened to a peg driven in the ground, which, as I have said before, should be put a sufficient distance from the coop to allow the hen to take shelter in it in case of need. A hen pegged down in this manner will become perfectly accustomed to the circumstances, and will proceed to scratch for her chickens in a very few minutes.

The advantages to the young birds from being on fresh, sweet ground, and obtaining natural food cannot be overstated. The hen commands a sufficient space of ground to prevent it becoming soiled, and she can be shifted day after day as often as required."

In situations where such a convenience is available, there is no more advantageous situation for newly hatched pheasants than a garden surrounded with high walls. A very practical correspondent, writing from Kildare, says: "There can be no better place to put young birds when newly reared than a large walled-in vegetable garden. I always place mine, hencoop and all, near a plot of cabbages, gooseberries, or raspberries, where they have good covert and feeding, and, above all, are protected from any injury at night during the period of their jugging on the ground, which they do for some time before they fly up to roost. By feeding them at the coops four or five times a day, they will stay in the garden until fully feathered, and able to fly over the wall to the adjacent coverts. I have had hen pheasants that nested in the garden and hatched under gooseberry bushes, coming to my whistle to feed regularly every morning. If the young birds are put out into the covert, the hen and coop (as in the garden) should be brought with them, and laid in a ride close to some very thick covert; they should be fed there about four times a day, beginning early in the morning, and diminishing as the birds grow strong. I feed them at this period on crushed wheat and barley, boiled potatoes chopped fine, some boiled rice and curds, all mixed together."

A very vexed question with regard to rearing of the young birds is the supply of water. Some very practical keepers give no water whatever; others give a very little; whilst a third set keep up an abundant supply. I am strongly of opinion that in this, as in all other respects, we cannot possibly do better than take nature for our guide. When hatched out naturally, there is no doubt that the birds obtain a plentiful supply of water. Even when there is no rain, the cloudless skies are productive of heavy dews, and the young birds may be seen drinking the glistening drops off the grass in the early morning. Some persons maintain that the ova of the gapeworm are taken in with the water gathered from dewdrops on the grass; others suggest that they occur in rain-water, but there is no foundation for either of these theories, as the disease is strictly local, which would not be the case if it were disseminated by a flying insect, by dew or rain water, or by any animals inhabiting running water. Much evil is produced by allowing the young pheasants to drink water contaminated with their own excrement, which is always the case if the water vessels are so constructed that the young can run into them; where such water is used, there can be no doubt of its injurious quality, but I cannot imagine that fresh, clear water can be otherwise than beneficial to the birds.

A correspondent, who is a most successful breeder of pheasants on a large scale, and whose young stock are in splendid order, writes: "I may give as my opinion that it is perfectly necessary to their health to have fresh spring water. Indeed, my man last year used to go to one particular spring to supply his birds, as it was better water. In their wild state, immediately they are out of the nest, the hen conducts them to the water, and in our wild Devonshire hills, where a streamlet runs in every valley, you can always see the well-defined paths of the broods to and from the water. I have just asked my man, and he tells me that so well are their water-loving propensities known, that poachers in large breeding places always net in dry weather any springs within reach of the coops, and often with success." Another authority says: "I am strongly opposed to attempting to rear pheasants without water, as against all nature; but my keeper adheres to his own opinion that for at least some weeks they should have it only once a day, bringing forward cases of broods hatched in dry fields where no water flows.

My idea is that in a wild state they can wander in search of dew, and also feed upon more moist and natural food than the egg, meat, and herbs that are chopped for them when reared under hens. I am aware that it is quite a common practice amongst keepers to deprive the little birds of water, and I cannot but feel it to be a cruel as well as a mistaken one. I believe that dry food wants water to aid digestion; and when birds are kept all day in small wired inclosures in the full blaze of the sun, it seems to me that they must require water to keep them healthy; and I also think that if they have a little always in the pen, they will drink less than when only given to them once a day. I saw a brood last week that had only had water once, quite early in the morning; they were being fed again in the evening, but would eat nothing. I then ordered some water to see what they would do, and the little birds and the old hen went to it at once, and seemed as if they could never have enough." And a third, writing to me on the same object, states: "I have been a rearer of pheasants for nearly thirty years. I give mine an unlimited supply of water at all stages of their growth, and I consider that it would be great cruelty to withhold it from them. I do not consider broods brought up by their mothers in dry fields where no water is to be found at all to the point. How can our poor artificial food compare with the thousand and one varieties they find in nature, full both of nourishment and moisture, with which it is impossible for us to supply them in confinement. I quite endorse your suggestion as regards the great value of lettuce for pheasants. I have fed them for some years with it, and they are very fond of it."

On the other hand, many successful keepers do not give water, or only in very small quantity. One correspondent says: "I know a keeper who rears a great number of pheasants each year, and he does not give them water till they are seven or eight weeks old, at which age they begin to eat barley and corn, and require water to assist digestion. He says that pheasants in their wild state take the dew in the mornings, and only in very dry weather do the old hens take their broods to water. In very dry weather, when there is little or no dew, he sprinkles water twice a day on the grass, but never puts any down for them until the time before stated, and when he waters the hens he does not allow the pheasants to drink." The writer of the following letter holds the balance very fairly between the opposing views: " Much depends on the nature of the food upon which the chicks are fed as to whether they should have water or not; if they are fed on dry food, and the weather is warm and dry, they will require water, but it must be very clean, and given only once a day, and must not remain before them longer than to allow each bird to have a little. If the birds are fed on moist scalded food, they will not require any water unless the weather is very hot, when a little may be given as before stated. The water must be spring or steam water, and I should advise it being given at noon. It must also be remembered that birds reared on heavy clay land will require less water than those reared on sandy or gravel soil; attention must also be paid to the amount of dew which falls, supposing the birds are set at liberty before the dew has time to evaporate. Those who argue that nature should be the guide on this point must recollect that the rearing of pheasants by hand is altogether an artificial process, and that therefore nature cannot be strictly followed with regard to water any more than with regard to food." A well-known game preserver writes on the subject as follows: " My keeper is a very successful breeder and rearer of pheasants. It seems to me (for I watched his proceedings very closely) that he gives the birds the very smallest supply of water. He carries a bottle in his pocket when he feeds, and puts about a wineglassful into each hen's saucer. The hens seem thirsty enough, and leave but little for the young birds. He feeds very sparingly but frequently, throwing the food wide. The food for a long time was rice with chopped boiled egg, ants' eggs, and a very few gentles. He has brought up a great many pheasants and birds for me. One year, strange to say, out of 211 he did not lose one. Certainly the season was favourable. Little water, and food thrown wide round the coops, seems to be his system." The scattering the food on clean soil being the most probable source of his success.

Inquiry is frequently made as to the cost of rearing pheasants in numbers. It is very difficult to state even an approximate sum, so much depends on the conditions under which they are raised. For food only until they are ready to go into the coverts, an average amount of from 1s. to 1s. 6d. per head may be stated. Mr. T. C. Cade writes : " The result of my own observations in two years (1870 and 1878) is as follows—In 1870 my keeper's bill for four hundred birds was, eggs, £5 6s. 6d. ; bread, £1 12s. 4½d. ; milk, £2 11s. 8d. ; suet, 13s. 6d. ; ' secrets,' 7s. 6d.—£10 11s. 6½d. To this must be added Indian corn, meal, and rabbits ; but I cannot give the exact quantity of each, as dogs were fed from the meal barrel, and the rabbits were not counted ; £9 8s. is, I consider, a fair estimate of the cost of what was used for the birds—making a total of £20 for four hundred, or 1s. each. About the ' secrets ' I can say nothing, except that none are required.

" In 1878, for three hundred under my supervision, the cost was : Very coarse Scotch meal, £9 15s. 6d. ; milk, £3 ; eggs, £1 15s. ; rennet, 2s. 8d. ; wheat, 8s. 8d. ; bread, 5s. 11½d. ; sheep's paunches (two hundred), £1 5s. ; a horse, 10s. ; a cow, 5s. ; a sheep, 5s. The last three for producing maggots. Total—£17 12s. 9½d. No rabbits were used. With this supply of food, at the cost of a little more than 1s. a head, not only were 97 per cent. of the birds reared, but I think they were as fine as possible."

The cost of labour, protection, &c., varies so much in different localities and under different circumstances that it is impracticable to draw up even a rough average of general application. Under very favourable conditions, as Mr. Cade demonstrates, pheasants may be reared at as small a cost as 1s. a head, and in others the cost rises to 14s. or even 15s.

The vignette at the end of this chapter represents the head of a partridge with a perfectly formed claw growing from the loose skin in the centre of the lower jaw; owing to the kindness of Mr. A. H. Stokes-Roberts I had the opportunity of examining the bird shortly after it had been shot. The drawing was made from the specimen in a recent state.

CHAPTER IX.

THE DISEASES OF PHEASANTS.

HEASANTS in a state of nature are particularly hardy. Being bred, as they generally are, from strong healthy parents, the few weakly chickens that are produced die under that benevolent arrangement which has been so justly termed the survival of the fittest in the struggle for life. Consequently the most vigorous remain as brood stock, and propagate a healthy offspring. Nevertheless, in some seasons, particularly during those that are wet, the young birds are affected by certain epidemic diseases that are difficult either to prevent or cure; amongst the first of these may be mentioned cold or catarrh, which is generally caused by an undue amount of wet weather acting on birds enfeebled by too close interbreeding, or by errors in the dietary and general management, such as undue exposure to cold winds. All that can be recommended in case of the young birds being thus afflicted is warm, dry shelter, and the addition of a little stimulating food, as bread soaked in ale, and spiced with any ordinary condiment, such as cayenne or common pepper, and the moistening of the oatmeal, or other soft food, with a solution of a quarter of an ounce of sulphate of iron in a quart of water, using enough to give the meal an inky taste.

Cold often runs on to roup, in which the discharge from the nostrils becomes purulent and infectious; in this case, the best mode of treatment is to endeavour to stamp out the disease by removing and destroying the affected birds instantly, and so preventing their affecting others. In all cases in which birds are destroyed to prevent the spread of any infectious disease, the greatest care should be taken not to leave the dead bodies exposed. If this be done, the disease is almost certain to extend; it has been proved to do so in the case of birds dying with tuberculous livers, "gapes," and other diseases. The bodies should, if possible, always be burned. If that is impracticable, they should be buried deeply in some part of the ground where there are no fowls or pheasants. Hanging the dead bodies of diseased birds in trees to produce a supply of gentles is exceedingly objectionable.

Scrofulous diseases, such as tubercles in the lungs and liver, are the result of breeding from weak stock, from overcrowding on the same ground, and from close interbreeding. The remedies suggest themselves; all that is required is the employment of strong, healthy stock birds, the removal to fresh untainted ground, and, if necessary, an introduction of fresh blood into the aviary or preserves.

The most troublesome and fatal disease is that known as the "gapes," which is caused by the presence of small red worms in the trachea or windpipe. For the first careful demonstration of the cause of this disease we are indebted to the late Dr. Spencer Cobbold, who contributed the account of its history and treatment to the Linnæan Society, from which the following abstract is taken :

"This parasite," writes Dr. Cobbold, "has been found in the trachea of the following birds, namely, the turkey, domestic fowl, pheasant, partridge, duck, lapwing, black stork, magpie, hooded crow, green woodpecker, starling, sparrow, martin, linnet, crow, rook, and swift.

"My attention was recently directed to a small, almost featherless chicken suffering from the 'gapes.' The bird belonged to a brood between six and seven weeks old. The

GAPEWORM IN PHEASANTS.

healthy birds had attained considerable size, and averaged 9½ ounces; the infested chicken weighed only 4 ounces; but, as if to make up for its defective assimilating powers, greedily devoured everything which came in its way, consuming two or three times as much as any other member of the brood."

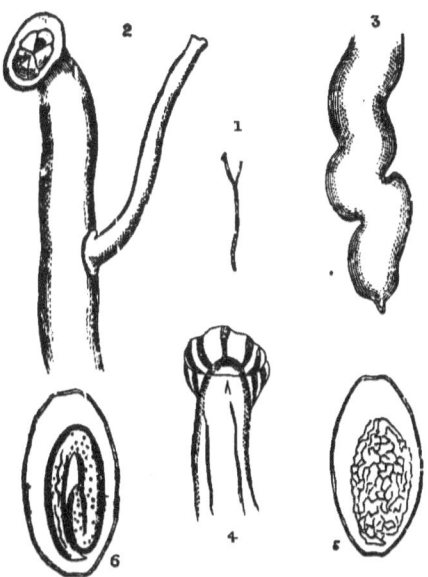

Fig. 1. *Syngamus trachealis*, male and female. Natural size.
Fig. 2. Upper part of the same, showing more especially the six-lobed circular lip of the female, and the mode of union. Enlarged.
Fig. 3. Lower end of the body of the female, with its mucronate caudal appendage. Enlarged.
Fig. 4. Lower end of the body of the male, showing the cup-shaped bursa, hard rays, lateral muscles, digestive tube, and round tail. Magnified 30 diameters.
Fig. 5. Mature egg. Magnified 220 diameters.
Fig. 6. Egg, with contained embryo. Magnified 220 diameters.

"The female worms extracted from the trachea have an average length of ⅝ths of an inch, the males scarcely exceeding ⅓th of an inch. In both sexes the bodies are tolerably uniform in breadth throughout. The mouth of the female is furnished with six prominent chitinous lips (Fig. 2). The male is usually found fixed by means of a strong membranous sucker (Fig. 4). The eggs of *Syngamus* are comparatively large, measuring, longitudinally, as much as the 1-250th of an inch (Fig. 5). Many of the ova contain fully-formed embryos. By whatever mode the young make their *exit* from the shell, it is manifest that prior to their expulsion they are sufficiently developed to undertake an active migration. Their next habitation may occur within the body of certain insect larvæ or even small land mollusks; but I think it more likely that they either enter the substance of vegetable matters or bury themselves in the soil at a short distance from the surface."

Since the publication of this paper, the history of the gapeworm has been very carefully studied by other observers, whose investigations have been recapitulated in Theobald's "Parasitic Diseases of Poultry." Mr. Theobald describes the ova and embryos as escaping by the rupturing of the female's body, which takes place, as a rule, after the worm has been expectorated by the fowl or pheasant. Both eggs and embryos, as suggested by Dr. Cobbold, take up their abode in damp ground, that around the drinking vessels of the fowls being a favourite locality.

The eggs hatch in from seven to forty days, according as to whether the surroundings are favourable or not. These ova and embryos get taken up by the young birds either off the ground or in the water; they then develop into the worms in the tracheal region of the fowl. The small embryo worms grow rapidly, soon become mature, and the females unite permanently to the males. They are often spread by one bird devouring the worms coughed up by another, and they are conveyed from one area to another by being carried by such wild birds as the starling and magpie, which are both very largely infested with *Syngamus trachealis*.

The theory that an intermediate host is necessary for the hatching of the ova has now been entirely disproved. Young fowls and pheasants quickly contract the disease when fed on contaminated soil in which the ova are present. Breeding pheasants on the same ground, as most gamekeepers know, constantly leads to "gapes," and direct experiment has shown that the disease may be introduced in healthy chickens by feeding them with the worms or ova. The theory of Dr. Walker that was published at length in *Nature* of August 2nd, 1888, by Lord Walsingham, that the eggs were hatched in the bodies of earthworms which are eaten by the young pheasants or fowls, has been entirely disproved by more carefully conducted experiments, as "gapes" appear in fowls on land where earthworms do not exist, and birds, such as the woodpecker, martin, and others, suffer from this disease though they do not eat earthworms.

With regard to the treatment of this disease, the plan of giving remedies internally to remove the worms is objectionable, as the medicine has to be absorbed, pass into the blood, and act powerfully upon the body of the bird before its purpose can be accomplished; its direct application to the worms is therefore preferable. This may be accomplished by stripping the vane from a small quill feather, except half an inch at its extremity; this should then be dipped in a mixture of one part of oil of turpentine and two of olive oil; and the chick being securely held by an assistant, the tongue may be drawn forward by catching the barbs at its base in a lock of cotton wool, and then pulling it forward so as to expose the small opening of the windpipe, down which the feather is to be passed sufficiently far to come into contact with the worms, and then turned round between the thumb and finger.

The application at once kills the parasites, and its application excites a fit of coughing, during which they are expelled: this mode of application requires some manual dexterity, and at times the irritation proves fatal; olive oil in the place of turpentine is sometimes employed.

Removing the worms by a feather is troublesome, and the operation is not always successful. Fumigation with

K

tobacco smoke is rarely of much avail. The administration of turpentine or camphor is attended with danger to the chickens, and opening the windpipe and extracting the worms whilst the bird is under the influence of chloroform requires surgical skill.

Knowing the active influence of carbolic acid on the lower forms of animal life, I determined to try the effect of the inhalation of its vapour in the cases of "gapes" that came under my notice. I have operated several times on chickens and turkeys that were suffering severely from "gapes," being almost choked by the worms. Each bird was placed in a small deal box, the open top being covered with a cloth. I then took a carbolic acid fumigator, consisting of a small metal saucer, heated by a spirit lamp. On the saucer I placed about a dozen drops of carbolic acid, lit the lamp, and put the apparatus in the interior of the box. Dense white fumes soon filled the box, and, being of necessity respired by the bird, came at once into contact with the worms. The operation was continued in every case until the birds were in some danger of suffocation. They soon, however, recovered on exposure to the air, and on the day following the treatment were running about perfectly free from any symptom of disease.

No special apparatus is required, as any arrangement which will serve to volatilise a few drops of the acid will answer; the vapour of carbolic acid may be used by putting a hot brick into the box, and pouring a few drops of the acid upon it, or it may be volatilised by putting three or four drops in a spoon, holding the latter over the flame of a lamp, and placing the head of the bird in the cloud of rising vapour. I have had a good deal of experience with birds afflicted with "gapes," but have never found any treatment equal to that of fumigation with carbolic acid vapour.

In very urgent cases, when the disease has so far advanced that immediate suffocation becomes inevitable, the opening of the windpipe, as adopted by Dr. Cobbold, may be advantageously had recourse to; or it may be resorted to when other methods have failed. In the most far-gone cases, instant relief will follow this operation, since the trachea may with certainty be cleared of all obstructions, but unfortunately it requires some amount of medical and surgical skill to administer the chloroform and perform the operation.

The most essential thing in view of putting a check upon the prevalence of the disease is the total destruction of the parasites. If the infected birds be thrown away, say upon the ground, the mature eggs in the gapeworms will not have sustained any injury. Decomposition having set in, the young embryos will sooner or later escape, migrate into the soil or elsewhere, and ultimately find their way into the air-passages of birds in the same manner as their parents did before them. The diseased birds ought to be burnt, and the dead bodies of any chickens, young partridges, or other birds infested with these parasites must be treated in the same manner if we wish to avoid the spread of the disease.

Since the publication of the last edition of this book, some exceedingly important investigations into the nature of the diseases of young pheasants have been made by Dr. E. Klein. The first of these diseases is that known to keepers under the name of "the cramps." This occasionally causes great mortality amongst young birds, attacking them usually during the second or third week. It is described by Dr. Klein as commencing with lameness in one leg. The next day the other becomes lame, and the bird sits motionless, and when made to move drags both limbs along the ground. Death generally occurs on the third day. On examination after death, the thigh-bone (the femur), or that of the leg (the tibia), or both, will be found soft, and in advanced cases broken, sometimes with great extravasation of blood into the surrounding tissues. This fracture generally occurs near the ends of the bone, whether that of the thigh or the leg. Microscopic examination shows that the interior of the bone is highly inflamed, the result of the presence of bacilli, which,

K 2

as in other infectious diseases, can be spread from one bird to another. The treatment of this disease is very simple. The moment it is recognised the young birds should be destroyed and burned. When this is acted on, at the beginning of the epidemic, it may prevent its further infection; but it is possible that the microbe may exist in the ground, from which it finds an entrance into the system of the birds. This view is very possibly the case, as experienced keepers know that "cramps" occur when the coops are kept on damp soil, and that when removed to higher dry ground the disease dies out. In a subsequent communication to the *Field* in the following year, Dr. Klein says: "I still attribute the cramp disease to a bacilliary infection of the system of the bird leading to corrosion and fracture of the bones."

Other epidemic infectious diseases affect young pheasants, carrying them off at times in large numbers. The most important of these is one, the origin of which is generally unsuspected amongst pheasant rearers, it arising from the farmyard hens which are used as hatchers and foster-mothers.

Dr. Klein gave a very careful account of this disease in the columns of the *Field*. He wrote as follows:

"I had the opportunity of investigating the disease in one of the eastern counties, where on one estate several hundreds of young pheasants became affected and died. The symptoms are these: The young birds, generally less than six weeks old, show either at one or both the angles of the mouth, on one or both eyelids, on the feet, sometimes also on the abdomen, some patches of various sizes and outlines, at first red and slightly elevated or swollen, then becoming yellowish-grey and dry and necrotic. When the eyelids are involved (which is the case in a large percentage) the birds appear blind, owing to the lids being more or less closed; where the feet are also affected (which in a large percentage is the case) the birds are weak and slow in walking, they limp also. When the mouth is affected they cannot feed, and therefore waste and soon die; the same result occurs when the eyelids become closed by the disease. In the large majority of fatal cases the affection involves one or both eyelids and the mouth; but in these cases also one or both legs show the disease in numerous necrotic patches of the skin. The disease is a cutaneous affection, and does not involve the deeper parts; on the legs the bones are unaltered, and there is no distinct visceral disease anywhere to be discovered by the naked eye inspection. Under the microscope in the earlier stages, the true skin is much inflamed, its vessels much congested, and the blood in them *in stasis;* the tissue of the skin is much infiltrated with inflammatory cells. Soon the whole inflamed parts begin to break down into a necrotic *débris*; the area of necrosis gradually enlarges, but is always surrounded by primary inflammatory change. This affection is therefore a true progressive necrosis of the skin.

" From a careful investigation, there can be no doubt that the disease is contagious, and further, that the first cases of disease amongst the young pheasants are due to infection by the same necrotic disease of the hens used for rearing. It ought to be stated that hens—fowls in general—are subject to, and not unfrequently affected with, an infectious disease, which shows itself as necrotic degeneration of the mucous membrane of the mouth and throat, and also of the skin around the mouth occasionally, but not often, also of the abdomen and chest. From inquiries which I instituted among the keepers, I feel convinced that in this particular locality of Suffolk the disease amongst the young pheasants was thus introduced, viz., by some diseased hens used for the rearing. It is obvious that if one hen is affected with the disease, the little pheasants that she is rearing are sure to contract it also, and these when affected, soon, in their turn, scatter the contagium over different parts of the field. When a hen is affected on the skin of the abdomen and chest, or when she has the disease in the mouth, sufficient of the contagium becomes available for the infection of the whole of her brood, which during the first weeks she is habitually covering with her body. From this it follows that the means to be adopted in order to exclude the disease from the pheasants at the outset are very simple, viz., carefully select the hens for rearing. They must be thoroughly examined before the hatching of the pheasants commences. The mouth and throat particularly, the skin of the abdomen and chest, must be healthy; where there is a sign of cutaneous necrotic disease, easily distinguishable as thick dry greyish-yellow friable deposits, the hen must be rejected. I know from inquiry that nothing in the shape of a careful selection actually occurs. Keepers take the hens wherever they can get them; they borrow them, buy them anywhere, or breed them. Sometimes they have the disease amongst their own poultry stock; but there is no attention paid to the healthy condition of the hens selected for rearing purposes. Apart from the losses amongst the pheasants by the disease, the fact that this disease is not uncommon amongst fowls, causes in some farms, considerable losses amongst the poultry itself. There is only one way of getting rid of the disease—that is, stamping out.

"When once an animal—be it fowl or pheasant—shows signs of the disease, it ought to be safely removed. When in any field where pheasants are reared the disease has made its appearance amongst the young birds, the hens ought to be carefully inspected, and the diseased hens and diseased pheasants removed. Those that are not affected ought to be placed on new ground. A field where the disease has been rife should not be used again for a year or two, and care should be taken that some disinfection be undertaken—*e.g.,* quicklime scattered over the field. But I feel sure that, if at the outset no diseased hen is admitted for the rearing, the disease will not make its appearance amongst the pheasants; for the hens seem to me to be the prime cause."

We are also indebted to Dr. Klein for the first accurate description of a very fatal epidemic disease which attacks fowls in overcrowded poultry-runs, and from them is apt to extend to pheasant coverts. This disease is termed by Dr. Klein *fowl enteritis*, or the "Orpington disease," inasmuch as "one well-known dealer had on his poultry farm, then at Orpington, in Kent, in about two acres of land, a fatal epidemic of fowls, by which he lost, between March, 1888, and March, 1889, over 400 birds."

He further states the disease to be highly infectious, as the evacuations of the diseased fowls are scattered about on the ground, contaminating the food which is picked up by the others, and rapidly spreads amongst the entire flock. The symptoms are severe purging of yellow evacuations, and the fowl is found dead in one or two days. The disease can only be checked by the immediate removal of the uninfected birds from the tainted ground, which should be disinfected with quicklime, or still better, gaslime, and well turned over. Every infected fowl should be at once taken away and destroyed, and the body burnt, not thrown on the ground, where the germs of the disease (bacilli) can spread. There should be no attempt at treatment even of the most valuable birds, and no chickens should be reared nor fresh stock placed on the tainted soil.

Some time since I received with a dead pheasant the following letter, showing how readily this fatal epidemic may spread from an overcrowded poultry-run into the coverts. The writer says:

"I am sending you with this a young pheasant which has been attacked with a disease that has unfortunately destroyed a large number of birds which were placed in the woods in a perfectly healthy condition. It is the general opinion that the birds have been affected by a poultry farm which is on the estate, as the fowls were known to be dying in large quantities from a similar disease."

On examination I found this bird affected with every symptom of fowl enteritis. The intestines showed redness in the mucous membrane, in the cæcal appendages there was a great amount of mucus, the spleen and liver were enlarged, and there is no doubt that the bacteria, or microbes causing the disease, could have been cultivated if it had been thought necessary to do so. There cannot be the slightest doubt that the disease affecting these pheasants was contracted from the fowls on the poultry farm on the estate, where they were dying in large quantities. The writer asks for a remedy. The researches of Dr. Klein, and the experience of those who have endeavoured to rear large numbers of pheasants or poultry on tainted ground, point to but one remedy, the destruction of the affected birds; and as it would be impossible to destroy the bacilli in the tainted ground over a large extent of covert, the rearing of pheasants should only take place on fresh and untainted ground the following year.

It is important to note that this fowl enteritis infects other gallinaceous birds, and that, as in the present case, pheasants in overcrowded pens and those reared in the neighbourhood of crowded poultry farms are liable to be attacked with the disease.

The moral to be drawn from these valuable researches of Dr. Klein is obvious. These infectious diseases are spread by the endeavour to rear pheasants and fowls on overcrowded and consequently tainted ground.

The remedy is the destruction by cremation of all the infected stock, the removal of those that are not diseased, and above all, the rearing of the poults and fresh pure ground.

Pheasants hatched under farmyard hens are not unfrequently liable to what are known as scurfy legs. The description of this objectionable disorder I may quote from my volume on "Table Poultry":

"Scurfy legs depend on the presence of minute parasites (*Sarcoptes mutans*), which live under the scales of the legs and upper part of the toes, where they set up an irritation, causing the formation of a white, powdery matter, that raises the scales and forms rough crusts, which sometimes become very large. When these crusts are broken off and examined with a microscope, or even a good hand lens, they will be found to be filled with the female parasites, generally distended with eggs. The crust itself may be compared to the crumb of dry bread; but the parasites are to be found only in those parts which are kept moist by the skin. They appear to cause great irritation to the bird.

"This disease is propagated by infection. It is seen in fanciers' yards where the poultry are closely confined together. The disease has been found affecting turkeys, pheasants, partridges, and even small birds in aviaries.

"The treatment in fowls is very simple. The legs may be soaked in warm water, and the crusts removed, and the legs washed with carbolic soft soap, as made for dogs; and the

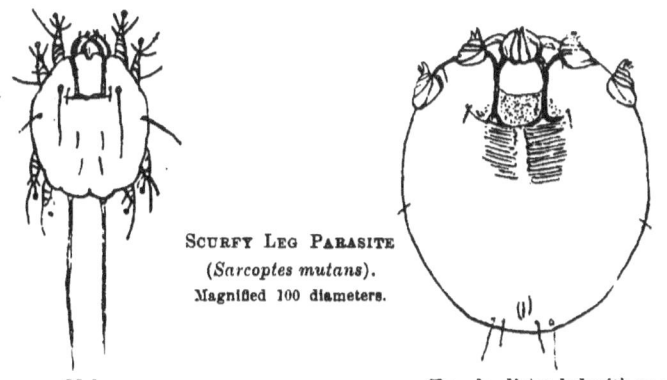

SCURFY LEG PARASITE
(*Sarcoptes mutans*).
Magnified 100 diameters.

Male.   Female, distended with eggs.

coops, nesting-places, perches, all cleansed with limewash, scented with carbolic acid. Great care should be taken not to employ as mothers any hens affected with the disease. If a Cochin or other hen in the slightest degree affected with scabies is employed, it is obvious that, as young birds are covered by her, the parasites can readily pass from her to the chicken, and the disease becomes disseminated."

The late Mr. Horne, of Hereford, a most practical pheasant rearer, wrote a letter to me on the subject, in which he states:

"There is no doubt that birds hatched under Asiatic

mothers are most prone to these insects. I have tried sulphur ointment, vaseline, glycerine, &c., but none were certain cures. At last I was told that common paraffin would speedily effect a cure. At that time I had a young bird (six months old) a perfect cripple—knots on his joints like nuts. I at once applied the paraffin, pouring it well over the legs; in a week there was a great improvement, and after two or three applications the bird became perfectly well. Since that time I have cured many. I generally apply it once in a week or ten days. I find the Versicolors and Reeves are the most liable to the disease, and do not remember having ever seen a case of it on the Gold."

Disease of the ovary, attended by the assumption of male plumage by the female pheasant, is a phenomenon that has long attracted the attention of naturalists. It was described by John Hunter in his "Animal Economy," and in the "Philosophical Transactions," vol. lxx., p. 527, and also by the late Mr. Yarrell. Although gamekeepers frequently speak of the hens thus changed in attire under the title of mule birds, it is now perfectly well known that the assumption of male plumage is invariably caused by disease of the ovary, and the birds exhibiting this change are, without any exception, always barren and useless females, not, however, necessarily old birds, as the change of plumage may result from ovarian disease in a hen that has not laid. The change takes place to a varying extent, usually beginning with a slight alteration of the neck feathers. In some cases it is absolutely entire; the hen being clothed in perfect masculine plumage, not a single feather of the body remaining unchanged. This singular modification is not confined to the common pheasant, but extends doubtless to the whole group. It is recorded as occurring in the Silver Pheasant (*Euplocamus nycthemerus*) in the *Field* of Nov. 13, 1869, and, thanks to the kindness of Mr. Leno, I had in my possession a Golden Pheasant hen (*Thaumalea picta*) in which the metamorphosis was complete. Mr. Leno had had this bird in his possession for some years, and had noticed the alteration increasing at each annual moult. A corresponding alteration has been frequently observed in the female of the domestic fowl, and it is not even confined to gallinaceous birds, being not unfrequent in the domestic duck. That disease of the ovary should cause the formation of feathers totally distinct, not only in colour, but in form, from those previously produced (as is most conspicuously the case of the tippet of the Golden, or tail of the Silver Pheasant) is a very remarkable circumstance, and one that has not yet received a satisfactory physiological explanation.

It not unfrequently happens that large numbers of young pheasants die of mysterious ailments, the causes of which are very difficult to determine. When they have been ascertained, they have not unfrequently been traced to some injurious substances that have been taken as food. In one case that came under my notice, the destructive agent was sheep's wool. A correspondent wrote, stating that during six weeks he lost upwards of 300 young pheasants from no apparent cause, but that subsequently he received a letter from his gamekeeper, who wrote :—" I have found out the cause of the pheasants dying. The farmer kept his sheep so long upon that piece of ground before I had the use of it, that the sheep lost a lot of wool, and my young birds have swallowed it. I have opened forty or fifty young birds, and found the gizzards quite full of wool, and the passage stopped up, so that food could not pass. I send you four pieces of wool, which I have taken from the gizzards of four different birds. I never had a better lot of young birds. They hatched off strong and well, and now I have lost nearly all of them."

It is probable that the sheep might have been dressed with some arsenical or other poisonous "dip" or "wash," which would remain on the wool and prove fatal to the young birds. The arsenical solution known as "weed-killer" is not unfrequently fatal to pheasants in pleasure grounds; it kills the worms and grubs that are near the surface of the paths, and these are eaten by the pheasants with fatal effect.

With regard to injurious substances taken as food, it is unquestionable that pheasants are sometimes destroyed by eating yew; but it is singular that the precise conditions under which they are poisoned have not been ascertained. The poisoning of animals from eating these leaves is so well known that damages have been claimed and obtained, after an appeal to the higher courts, by persons who have lost cattle, horses, or sheep, in consequence of the branches of yew trees being allowed to hang over fences, or the cutting of hedges being thrown upon the ground. In conjunction with the late Professor Tuson, of the Veterinary College, I investigated the poisoning of pheasants by yew leaves several years ago. The action of the poisonous leaves in producing inflammation of the intestines was so well marked that there could be no possible doubt of the cause of death; but the circumstances that lead well-fed pheasants to eat yew leaves on some occasions, and not to touch them on others, are difficult of explanation. The poisoned birds that I have examined have always been highly nourished, extremely fat, and in good condition, and, so far from being hungry, their crops in many instances have been filled with maize.

I have recently received one of several pheasants that had been picked up dead in the coverts of Mr. Ryde, of Chiddingfold. This pheasant was in the most splendid condition; the crop contained about a dozen leaves of yew and a few grains of small maize. There were also comminuted leaves in the gizzard, and distinct evidence of their existence in the intestines, and there could be no doubt of the cause of death.

Some few years ago Lieut. F. Stuart Wortley, then working at the Agricultural College, Downton, wrote a letter to the *Times* in which he described a number of experiments performed with a view of ascertaining the amount of the poisonous principle known as taxine in the leaves of the male and female yew respectively. His experiments definitely proved that taxine exists in a much larger quantity in the leaves of the male than in those of the female yew. If this taxine is the active principle, his experiments tend to prove that only the male yew is poisonous, but I am not aware whether any further experiments have been since made on the subject. It would be very desirable that some observer who has the opportunity should ascertain by actual experiments whether there is any difference in the action of the leaves of the male and those of the female yew when given to pheasants or other animals. This could be readily accomplished by mixing the leaves of the two trees with ground meal, and administering it to pheasants in captivity. The information thus obtained would be very valuable, inasmuch as if it were found that the leaves of the female yew were not poisonous, it would lead to their being safely planted in coverts and places accessible to animals. A great deal of the doubt and uncertainty which prevails respecting the poisoning of animals by yew may possibly depend upon the relative amount of poison contained in the leaves of the two sexes of this plant. It is well known that children often gather and eat the waxy covering of the berries of the yew without injury, consequently in that part of the plant there can be no amount of this bitter principle known as taxine. The whole matter requires a little more careful investigation, and offers a very interesting subject of experiment to any person with the means at his disposal.

Another frequently unsuspected cause of death in pheasants is the habit they acquire of picking up and swallowing shot when in coverts that are much shot over. Mr. J. Hindle Calvert, F.C.S., made the following communication to the *Field* in 1876, and his inferences have been since amply confirmed by myself and others who have made post-mortem examinations in similar cases. Mr. Calvert wrote:—" The following cases of lead poisoning in pheasants may be of interest to those who have large pheasant preserves. A gamekeeper brought me for inspection a hen pheasant which was partially paralysed in the legs, and low in condition. On killing the same and opening the gizzard I found thirteen leaden pellets of various sizes; the grinding action of the gizzard had disseminated the lead with the food, and the bird was surely but safely undergoing the slow process of lead-poisoning. This was very evident on applying the usual chemical tests, as I readily detected lead dissolved in the food, and also traces in the blood taken from the region of the heart. Two days after this the gamekeeper brought another live bird. This one had been in a sickly condition for two or three weeks, and was quite emaciated. The legs were paralysed, and the feet drawn in a similar manner to the drop-hand, when lead has been the cause of poisoning in the human subject. On opening the gizzard I found four pellets, so that there is little doubt that this bird would soon have died from the effects of lead-poisoning.

"I understand last year some score of pheasants died in the same preserve, all of them showing symptoms same as above related. Both years the poisoning happened after the coverts had been shot through. No doubt the birds pick up the pellets under the delusion of being either food or grains of sand; perhaps the latter. When the birds died last year the cause of death was attributed to there being too many left for breeding purposes; rather a strange reason, seeing that the birds had been decimated on the shooting day.

"Others may have experienced something similar to the above, without being able to give a satisfactory reason for the birds dying; but where you have paralysed limbs and a gradual falling off in condition, and should this happen some weeks after the covert has been shot through, then they may suspect that lead-poisoning is a probable cause."

COMMON PHEASANT (*Phasianus colchicus*).

CHAPTER X.

PHEASANTS ADAPTED TO THE COVERT.

THE COMMON PHEASANT (*PHASIANUS COLCHICUS*).

HE pheasants which are best adapted to the coverts in England, the United States of America, Australia, and other temperate climates, are undoubtedly those which belong to the restricted genus *Phasianus*, or, as so many term them, the true pheasants. Formerly there was but one distinct species or race known in Europe, that which is named the *P. colchicus*, from its having being received from the banks of the River Colchis in Asia Minor. This was followed by the ring-necked *P. torquatus* from China, and subsequently by the *P. versicolor* from Japan. These were originally regarded by naturalists as perfectly distinct species, but it is now known that they breed freely with one another, and that the offspring are perfectly fertile, however intimately they are interbred. The late Henry Seebohm, who paid great attention to the birds of this group, writing in the *Ibis* for 1887, said:

"The fact that all true pheasants interbreed freely with each other and produce fertile offspring, may be accepted as absolute proof that they are only subspecifically distinct from each other. Like all other subspecies, they only exist upon sufferance. The local races appear to be distinct enough, but they only retain their distinctive character as long as they are isolated from each other. The moment they are brought into contact they begin to interbreed; crosses of every kind rapidly appear, and in a comparatively short time the swamping effects of interbreeding reduce the two or more local races which have been brought into contact to a single and uniform intermediate race. Such swamping effects of interbreeding have practically stamped out in the British Islands the two very different looking races of pheasants which were introduced into them——*Phasianus colchicus* from Asia Minor, and *Phasianus torquatus* from China. The pheasant of the British Islands is, with very rare exceptions, only a mongrel between these two races, but, it must be admitted, a very healthy and fertile one."

The intermingling of the several races in the course of ages, and the isolation of the different breeds in the valleys and river systems of Asia, have given rise to numerous sub-species which are found spread over that vast continent. The spread of scientific investigation is continually disclosing new pheasants, which it pleases the discoverers to regard as distinct species, but which are obviously only mixed races. Mr. D. G. Elliot, writing in 1872, enumerated about a dozen. Mr. Seebohm, in the *Ibis* for 1887, described six as sub-species of *P. colchicus* (three of which were not recognised by Mr. Elliot). These are *P. principalis* from North Afghanistan; *P. persicus*, which Mr. Elliot regards as the same as *P. shawi*; and *P. chrysomelas*, which he regards as identical with *P. insignis*. In the following volume (1888) Mr. Seebohm enumerates seven races, of which the Chinese *P. torquatus* may be regarded as the type; of these two, *P. vlangali* and *P. strauchi*, are not described by Elliot. Of the others, the most strongly marked is the Japanese *P. versicolor*, which appears to me to be the most distinct and typical of all the true pheasants.

It would be but a tedious and most unprofitable waste of time to enter into all the fifty so-called breeds of pheasants which the species-mongers have raised to the dignity of distinct species; suffice to say they are all perfectly fertile, *inter se*, as are their progeny to any extent. It may please closet naturalists to classify them, not knowing how easily they may be bred, and to give them specific names after their friends, which has been done in a dozen or more instances. But the naming a variety as a new species because it has a few feathers on the neck or wing whiter or darker than those of another, has little interest for practical men.

In the following pages the more typical species will be described, and their numerous varieties treated as allies.

In commencing the description of the different pheasants adapted to the covert, the common species (*Phasianus colchicus*) claims the first place, as it is more generally distributed and better known than any of the more recent introductions. Although not equalling some of them in size, or gorgeousness of plumage, it is by many sportsmen preferred in consequence of its rapid flight and active habits. It is, however, only in the remote districts of the country that it is now to be found in a state of purity, as the introduction of the Chinese and Japanese races has given rise to so many cross-bred varieties that in many places a purely bred *P. colchicus* is a rarity.

Lord Lilford, in "The Birds of Northamptonshire," writing of the common pheasant, says:—" Although it is now difficult to find pure-bred specimens of this species, on account of the frequent crossings with the Chinese Ring-necked Pheasant (*Phasianus torquatus*) and other species, we do occasionally meet with birds, especially in the large woodlands of the northern division of Northamptonshire, which, by their small size, the absence of any trace of the white collar, which is so conspicuous in the Chinese bird, and the intense blackness of the plumage of the lower belly, present the characteristics of the true unadulterated species."

In the district of the Humber we are informed by Mr. John Cordeaux that "the pure old breed untainted by any cross is now seldom to be met with, excepting in a few localities furthest removed from the great centres of game preserving. With these few exceptions, our resident birds are a mixed race, exhibiting in a greater or less degree the cross between the old English bird and the Ring-neck (*P. torquatus*)." This statement is equally true of all the well-preserved districts of England, in many of which the varieties are still more complex in consequence of the introduction of the Japanese species (*P. versicolor*).

Under these circumstances, I have thought it desirable to quote the description of the common pheasant from the first volume of Macgillivray's "British Birds," 1837, inasmuch as the author's descriptions are admirable for their accuracy and attention to detail, and at the date at which it was published the common species had not in Scotland been crossed with any of the more recent importations.

The following is Macgillivray's description of the two sexes of *Phasianus colchicus* :—

"Male.—The legs are stronger; the tarsi, which are stout and a little compressed, have about seventeen plates in each of their anterior series. The first toe, which is very small, has five, the second twelve, the third twenty-two, the fourth nineteen scutella. The spur on the back of the tarsus is conical, blunt, and about a quarter of an inch long.

"The feathers of the upper part of the head are oblong and blended, of the rest of the head and the upper part of the neck imbricated and rounded, of the fore-neck and breast broad, slightly emarginate or abruptly rounded; of the back broad and rounded, of the rump elongated, with loose filaments; of the sides very long, of the abdomen downy, of the legs soft and rather short. Directly over the aperture of the ear is a small erectile tuft of feathers. The wings are short, very broad, curved, rounded, of twenty-four quills; the primaries attenuated from near the base, rounded, the third and fourth longest, the first equal to the seventh; the secondaries broad, rounded, and little shorter than the primaries. The tail is very long, slightly arched, remarkably cuneate or tapering, of eighteen tapering feathers, of which the lateral are incurved, the central straight. Four pairs of the longest tail feathers are concave above towards the end, or channelled.

"The bill is pale greenish-yellow, the nasal membrane light brown or flesh-coloured. The bare papillar patch on the side of the head is scarlet, in parts approaching to arterial blood-red, or at some seasons crimson. The eyelids are flesh-coloured, the iris yellow. The feet are light grey tinged with brown, the claws light chocolate brown.

"The feathers of the upper part of the head are deep brownish-green, with yellowish marginal filaments. The upper part of the neck is deep green behind, laterally and anteriorly greenish-blue and purplish blue. The lower part of the neck is reddish-orange, anteriorly tinged with purple; the breast and sides brownish-yellow; each feather terminally margined with purplish-blue, the dark margin indented in the middle, but the indentation gradually diminishing on the breast. The middle of the lower part of the breast is blackish-brown, glossed with green, the margins of the feathers being of the latter colour. The fore part of the back is yellowish-red, each feather slightly margined with black, and having a central oblong spot of the same. The scapulars are redder, with a slight black tip, the central part dull yellow mottled with dusky, margined with a black band. On the middle of the back the feathers are somewhat similarly variegated, with additional spots of light blue and purple. Those on the rump are of a deep red, with green and greyish tints. The inner wing-coverts are similar to the scapulars, but edged externally with dark red, the outer yellowish-grey, variegated with whitish and dusky. The quills are light brownish-grey, variegated with pale greyish-yellow; the secondaries more tinged with brown on the outer edges. The tail is dull greenish-yellow, variegated with yellowish-grey, the feathers with narrow transverse bars of black, a broad longitudinal band of dull red on each side, the loose margins red, glossed with green and purple. On the abdomen and legs the feathers are dull greyish-brown; under the tail variegated with reddish. The lower surface of the wing is yellowish-grey.

"Length to end of tail 34 inches; extent of wings 32; wing from flexure 10; tail 18½; bill along the back $1\frac{4}{12}$, along the edge of upper mandible $1\frac{5}{12}$; tarsus $3\frac{2}{12}$; first toe $\frac{7}{12}$, its claw $\frac{3}{12}$; second toe $1\frac{1}{12}$, its claw $\frac{6}{12}$; third toe $2\frac{1}{4}$, its claw $\frac{8}{12}$; fourth toe $1\frac{8}{12}$, its claw 4½ twelfths.

"Of three other individuals, the length 34, 35, 36 inches.

"Female.—The female is similar in form to the male, but with the tail much shorter. The bill and feet require no particular description. The anterior scutella of the tarsus are about seventeen in each row; the first toe has five, the second fifteen, the third twenty-two, the fourth eighteen. As in the male, there is a bare space under the eye, but scarcely papillar, and more feathered. The feathers of the upper part of the head are somewhat elongated; those of the rest of the head short; of the neck and body oblong and rounded; of the rump not elongated as in the male.

"The general colour of the upper parts is greyish-yellow, variegated with black and yellowish-brown; the top of the head and the hind-neck tinged with red. The wing-coverts are lighter; the quills pale greyish-brown, mottled with greyish-yellow, as in the male. The tail is yellowish-grey, minutely mottled with black, and having in place of transverse bars, oblique irregular spots of black, centered with a pale yellow line. The lower parts are lighter and less mottled, the throat whitish, and without spots. The bill is horn-coloured, tinged with green; the tarsi wood-brown, the toes darker, the claws of the same tint.

"Length 26 inches; extent of wings 30; wing from flexure 9¼; tail 11½; bill along the back 1¼; tarsus 2¼; first toe ½, its claw $\frac{4}{12}$; second toe $1\frac{2}{12}$, its claw $\frac{6}{12}$; third toe $1\frac{10}{12}$, its claw $\frac{7}{12}$; fourth toe $1\frac{4}{12}$, its claw $\frac{5}{12}$."

Several well-marked and perfectly permanent varieties of

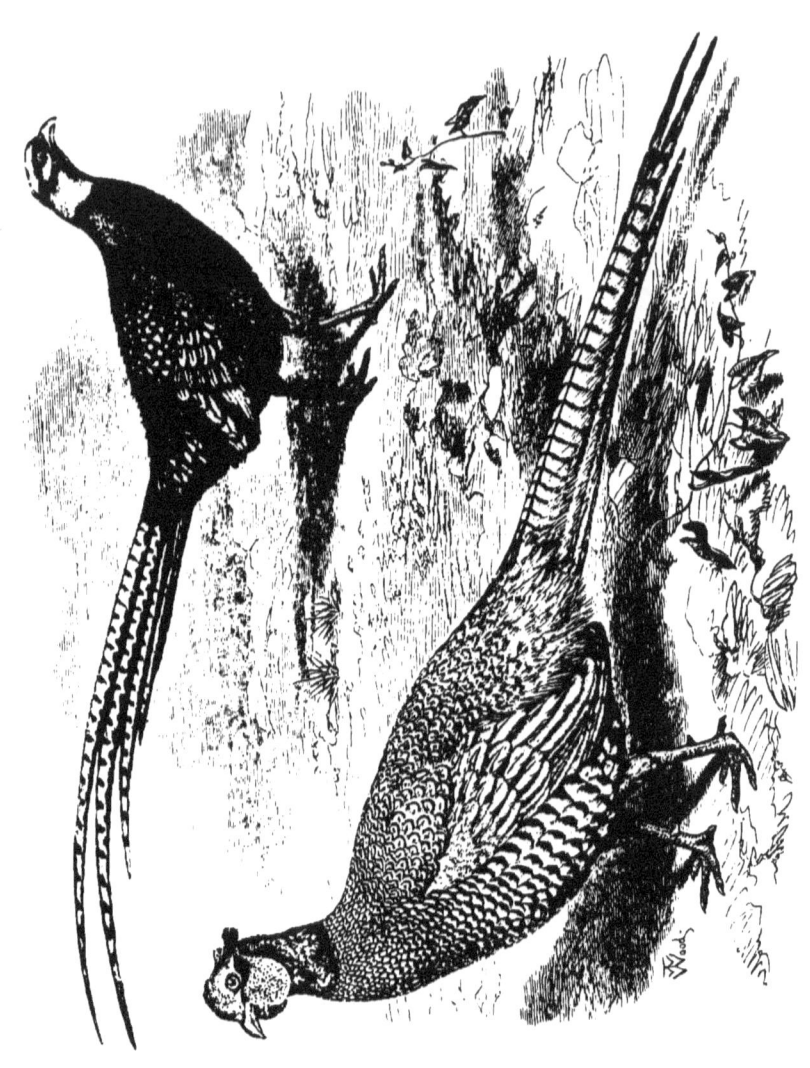

this species are not uncommon. One of the best known is the so-called Bohemian pheasant, in which the entire plumage is much less glossy, the general ground-colour being of a creamy tint; the head, neck, and spanglings on the breast and tail showing the dark markings in varying degrees of intensity in different specimens. The appearance of this variety is admirably given in the engraving. The Bohemian pheasant is occasionally produced from the common form in different localities, the variation is hereditary, and may be propagated by careful selection of brood stock. Thus Mr. Stevenson, in his "Birds of Norfolk," informs us that in that county, like certain light varieties of the common partridge, they are confined to particular localities :—"They have been found in different seasons in some coverts at Cranmer; and in the autumn of 1861, I saw three fine examples killed, I believe, in Mrs. Hardcastle's preserves at Hanworth, near Cromer, one of which, even in its abnormal plumage, showed a decided relationship to the Ring-necked cross, by the white mark on either side of the neck"—a circumstance also noticed by Macgillivray.

A purely white variety of the common pheasant occasionally occurs in the coverts without any apparent cause. A correspondent, who has been a pheasant rearer for thirty years, writes :—" Four years ago a nest of thirteen eggs was brought in by the mowers. All the eggs were hatched; eleven were perfectly white birds, the other two the common colour. Nine of the white birds were reared—six cocks and three hens; three cocks were turned out, the others were kept in the pheasantry, pinioned. The white pheasants proved very bad layers—very delicate, their eggs very bad; and those that were hatched very difficult to rear, and there never was a white bird bred. The extraordinary thing is, that where the nest was taken up the keepers had never before or since seen a white pheasant. The three cocks turned out never (to my knowledge or the keeper's) were the cause of white pheasants or pied pheasants being bred, and the three all disappeared in the second year. On another part of my estate a white cock pheasant was bred; he was considered a sacred bird, and lived seven years, when he disappeared. In the covert he resorted to I killed one pied pheasant, and I believe that one bird was the only pied pheasant (if bred through him) that ever was seen." By careful breeding there is no doubt that a permanent white race might be established if such a proceeding were thought desirable, which I much doubt, as white varieties are generally very deficient in hardihood. Left to themselves, the white cocks are doubtless driven away from the hens by the stronger and more vigorous dark birds, and rarely increase their kind. When mated in pheasantries the natural colour has a strong tendency to reproduce itself; but white, or even pied or parti-coloured birds, are not always to be produced from white parents, as the following letters will show:—"On the manor of a friend in Yorkshire are a cock and hen pheasant entirely and purely white. They inhabit different woods, and are strenuously protected by the head keeper, who considers their presence a proof of the integrity of his coverts, and invariably requests strangers to spare them. There are also a few ring-necks in the coverts, which have bred so freely with the common sort that hardly a cock pheasant is killed but shows some marks of white about his neck, while pied birds are so rare that the few that have been shot have been preserved. If, then, white pheasants breeding with ring-necks and other birds produced, as a rule, pied birds, why should there not have been every year at least one brood of pied pheasants in these woods in the same proportion as the half-bred ring-necks?" Another correspondent writes:—"A white hen was confined in the pheasantry here for some years with a common pheasant, but of the progeny there was not one pied bird. A pied cock was then confined with a common hen pheasant, and there were a few of the chicks pied. Lastly, a pied cock and a pied hen were confined together, and invariably every one of the chicks was pied. I have tried the experiment frequently with the same results." And a third states:—"I deny that the cross between the white and common pheasant will produce pied, when both are pure bred. I have tried the cross in confinement for years, and never produced one pied bird from it; and before the pied breed were introduced into the preserves here, we had an abundance of white cocks and white hens, and, believing at that time that the pied was the result of a cross between the white and common pheasant, I used to watch the nides of every white hen, and was surprised that in no instance was there one pied chick, though some were white."

The explanation of the difficulty of breeding pied birds from a white and a coloured parent, and the ease with which ring-necks are produced and perpetuated, is soon given. Ring-necks are derived more or less directly from the *P. torquatus*, a permanent race, that has a strong tendency to reproduce its like; but white and pied birds are merely accidental variations, and not even a thoroughly established breed, and therefore are not prepotent in propagating their like, but have a strong tendency to throw back to the original stock from which they were derived.

CHAPTER XI.

PHEASANTS ADAPTED TO THE COVERT (CONTINUED).

THE PRINCE OF WALES'S PHEASANT (*PHASIANUS PRINCIPALIS*).

HE pheasant most recently introduced in a living state into this country is that named after H.R.H. the Prince of Wales; the account of its introduction is soon told:

In April, 1885, Mr. Sclater exhibited at the Zoological Society skins of a pair of pheasants brought from Bala Murghab, North Afghanistan, belonging to the Prince of Wales, and read extracts relating to the specimens from a letter addressed by Mr. Condie Stephen to Sir Dighton Probyn.

"These pheasants," writes Mr. Stephen, "abound in the reeds fringing this river, rising in places in far larger numbers than I have seen at any battue in England. You can imagine what a quantity there must be from the fact that we killed more than four hundred on our march of thirty miles up the river, mostly cocks."

The living specimens, which were subsequently received in the gardens, and from which the engraving was taken, were obtained by Major Peacock from the Afghan frontier, but in consequence of their being received in very poor condition, they all died without having produced any young.

The most important characteristic of this fine bird, and one that distinguishes it from all those pheasants naturalised in this country, is that the wing coverts are white, a character which it has in common with the two pheasants named *Phasianus insignis* and *P. mongolicus*. It differs somewhat in the markings and arrangement of the colours from either of these birds, and has consequently been regarded as a distinct species, and named *Phasianus principalis*, in commemoration of the fact that the skins were received from H.R.H. the Prince of Wales.

If it should be successfully introduced—and there is no reason why it should not be—it will be a welcome addition to our coverts, giving size and hardihood to our native birds. The plumage is beautifully spangled with bright purplish black on a rich golden red ground, the white shoulders and dark flight feathers standing out in strong contrast; but there is no ring around the neck, as in the Chinese pheasant.

In its habits it differs somewhat from our common species in frequenting swampy ground covered with reeds, returning to the covert to roost at night. In its native habitat it is most abundant. At Masuchak, on the Upper Murghab, in Northern Afghanistan, Major Durand and Major Yate, as recorded in the latter officer's letters from the Afghan Boundary Commission," brought in a bag of nearly fifty pheasants (*Phasianus principalis*) killed during the afternoon. "It is extraordinary," Major Yate remarks, "what a number of pheasants there are in the reed swamps of this valley, and this year they seem to be even more numerous than last. I know of no country in the world where one can get such good real wild-pheasant shooting as this. On the 21st we also brought in a bag of seventy-two pheasants, but, as on the first day, lost a good many wounded birds. The reeds are so thick, and the birds, especially the old cocks, so strong, that it is very hard to bag one's bird even after it is shot."

Dr. Aitchison, writing of this pheasant in the transactions

of the Linnean Society, says: "The specimens of this pheasant were all got on the banks of the Bala Murghab, where it occurs in considerable numbers in the tamarisk and grass jungle growing in the bed of the river. More than four hundred were killed in the march of thirty miles up this river. It not only wades through the water in trying to make from one point of vantage to another, but swims, and seems to be quite at home in these thickets, where there is always water to the depth of two or three feet. These swampy localities afford good shelter. In the mornings and evenings the pheasants leave it for the more open and dry country, where they pick up their food. I believe the same species is found on the Hari-rud river, but I have seen no specimens from that locality."

It is not without interest to note that, though not yet bred in Europe, the Prince of Wales' pheasant has made its appearance in considerable numbers in the London markets, having been imported through Russia in a frozen state, and it is recorded that the late Mr. Seebohm, after having given £10 for one of the original skins received from Afghanistan, supplemented it the following year by a couple selected from a batch of birds in Leadenhall Market and bought for a few shillings.

The introduction of this pheasant into our coverts would be advantageous, not only on account of its size and plumage, but still more so from the fact that, having been reared on swampy ground, it would be a very desirable addition to our coverts in many localities.

CHINESE PHEASANT (*Phasianus torquatus*)

CHAPTER XII.

PHEASANTS ADAPTED TO THE COVERT (CONTINUED).

THE CHINESE PHEASANT (*PHASIANUS TORQUATUS*).

CONSUL SWINHOE, Mr. Dudley E. Saurin, Père David, Prjevalski, and other naturalists, who have investigated the fauna of the Chinese empire, unite in confirming the belief that this pheasant (*P. torquatus*) is the most common species in China, abounding in vast numbers in the hill coverts and cotton fields. Mr. Saurin states: " The common Chinese pheasant is found everywhere in the north of China. I am not aware how much further south they are found than Shanghai; but in that neighbourhood, since the devastation of the country by the Tai-pings, they are shot by hundreds. Thousands are brought down to the Pekin market in a frozen state by the Mongols, from as far north as the Amour. At the new Russian port of Poussiet, conterminous with the Corea, the same pheasant abounds. I myself have seen them wild in the Imperial hunting grounds north of Jehol, and in the mountains near Ku-peh-kow."

Consul Swinhoe says that it is very common near Hankow, and at all the places that have been visited by Europeans north of the Yangtze. Formosa swarms with these birds; the specimens found there, however, differ from those of the typical race by having the ochreous feathers on the flanks exceedingly pale, and by some writers this local variety has been described as a distinct species under the name of *P. formosanus*.

The characters of the pure-bred Chinese *P. torquatus* were given in minute detail by the late Mr. Gould, in his magnificent folio, "The Birds of Asia." They are as follows :—
"The male has the forehead deep green; crown of the head fawn colour, glossed with green; over each eye a conspicuous streak of buffy white; the naked papillated skin of the orbits and sides of the face deep scarlet or blood red, interspersed beneath the eye with a series of very minute black feathers; horn-like tufts on each side of the head; throat and neck rich deep, shining green, with violet reflections; near the base of the neck a conspicuous collar of shining white feathers, narrow before and behind, and broadly dilated at the sides; the feathers of the back of the neck black, with a narrow mark of white down the centre of the back portion, and a large lengthened mark of ochreous yellow within the edge of each web near the tip; the feathers of back and scapularies black at the base, with a streak of white in the middle, then buff surrounded with a distinct narrow band of black, to which succeeds an outer fringe of chesnut; feathers of the back black, with numerous zigzag and crescentic marks of buffy white; lower part of the back, rump, and upper tail coverts light green of various shades, passing into bluish-grey at the sides, below which is a mark of rufous; breast feathers indented at the tip, of a rich reddish chesnut, with purple reflections, and each bordered with black; flanks fine buff, with a large angular spot of beautiful violet at the tip; centre of the abdomen black, with violet reflections; under tail coverts reddish chesnut; wing coverts silvery-grey; wings brown, the primaries with light shafts, and crossed with narrow bars of light buff; the secondaries similar, but not so regularly marked as the primaries; tail feathers olive, fringed with

different shades of reddish violet, and crossed at regular intervals with broad, conspicuous black bands, passing into reddish-brown on the sides of the basal portion of the six central feathers; bill yellowish-horn colour; irides yellow; feet greyish-white. The female has the whole of the upper surface brownish-black, with a margin of buff to every feather; the throat whitish, and the central portion of the under surface fawn colour; flanks mottled with brown; tail buff, barred with dark brown, between which are other interrupted bars of the same hue. These marks are broader on the two central feathers than on the others, and, moreover, do not reveal the edge on either side."

The specific name *torquatus* is derived from *torquis*, a chain or collar worn around the neck. This species was introduced into England a great many years since, long before the time of Latham, who described it as having been turned out in preserves on many estates. No birds could be better adapted for our coverts; being natives of a cold part of China, they are very hardy—a character which they display by laying early in the season, and by producing an abundant supply of eggs. The pure Chinese is a bird of bold flight, rising through the covert with great quickness, and then pursuing a swift, straight course. It is unquestionably a most ornamental addition to our game birds, being valuable not only for the beauty of its plumage, but also for the delicacy of its flesh. The breed is, however, kept in a state of absolute purity with some difficulty, as the males are apt to wander to "fresh woods and pastures new." Hence crosses between it and the common species are very prevalent; these constitute what are usually called the ring-necked pheasants. These cross-bred birds are perfectly fertile, not only with either pure race, but also *inter se*. They are, however, variable in plumage, the amount of white in the neck varying from four or five feathers to a nearly complete circle, and the feathers on the flanks being intermediate between the beautiful spotted buff of the pure Chinese and the dark colour of the common bird.

These ring-necks are now common in most parts of the country where pheasants are preserved. The good points of the Chinese are largely shared by their half-bred progeny; hence the cross between the common and the Chinese is a valuable introduction to our preserves, retaining as it does to so great a degree the beauty and early fertility of the pure Chinese race, to which it adds great hardihood and larger size, but the birds are generally regarded as more apt to stray, and some gourmets maintain they are not quite so good a bird on the table as the pure-bred *P. colchicus*.

The extent to which the interbreeding of the two species has taken place is well shown in the following interesting account taken from Mr. Stevenson's "Birds of Norfolk":—
"In its semi-domesticated state, like our pigeons and poultry, the common pheasant crosses readily with its kindred species, and to so great an extent has this been carried in Norfolk that, except in the wholly unpreserved districts, it is difficult at the present time to find a perfect specimen of the old English type (*P. colchicus*) without some traces, however slight, of the ring-neck, and other marked features of the Chinese pheasant (*P. torquatus*), and in many localities of the Japanese (*P. versicolor*). In looking over a large number of pheasants from different coverts, as I have frequently done of late years in our fish market, I have noticed every shade of difference from the nearly pure-bred ring-neck, with its buff-coloured flanks and rich tints of lavender, and green on the wing and tail-coverts, to the common pheasant in its brilliant but less varied plumage, with but one feather in its glossy neck just tipped with a speck of white. Some birds of the first cross are scarcely distinguishable from the true *P. torquatus*, and are most gorgeous objects when flushed in the sunlight on open ground; but as the 'strain' gradually dies out, the green and lavender tints on the back begin to fade, and the rich orange flanks are toned down by degrees; though still the most marked feature of all, the white ring on the neck, descends from one generation to another, and the

hybrid origin of the bird is thus apparent long after every other trace of its mixed parentage has entirely passed away."

The Chinese pheasant has been introduced into several parts of the globe with success. The rapidity of its increase in New Zealand has already been noticed. As long since as the year 1513 it was acclimatised in the island of St. Helena under very peculiar circumstances, as related by Brookes in his history of the island. Fernandez Lopez, having deserted from the army of A. Albuquerque at Goa, was exiled, along with a number of negroes, and banished to St. Helena, being supplied with roots, seeds, poultry, and pheasants for turning out. These were of the species now under consideration. Berries and seeds being abundant in the island, the birds became wild, throve amazingly, and on the visit of Captain Cavendish in 1588 he found them in great abundance and admirable condition. In 1875 we are informed, in Melliss's "St. Helena," "that they still exist abundantly, and quite maintain the characteristics mentioned by Cavendish. They are protected by game laws, which permit them to be killed, on payment of the licences, for six weeks in the summer or autumn of each year, and hundreds of them are generally killed during one shooting season. They find plenty of covert, and generally make their nests in the long tufty fields of cow-grass (*Paspalum scrobiculatum*)."

There can be no doubt that the Chinese or ring-necked species has remained in its purity at St. Helena. Ships going to India *viâ* the Cape of Good Hope in the olden time did not sail within a thousand miles of St. Helena; but, taking advantage of the trade winds, they went direct to the coast of South America, often, indeed, calling at Rio, and then struck straight away for the Cape of Good Hope, aided by the return trade wind. It was on the return from India that the Island of St. Helena was visited, and letters from England to the island went *viâ* the Cape. Under these circumstances, the introduction of a Colchian pheasant to the island is exceedingly improbable, and that of a Japanese out of the question. With regard to the alteration in plumage produced by an exposure to these new conditions for 373 years, it must be confessed that they are remarkably insignificant. There is the same glossy, shining green of the head and neck, the white ring completely surrounding the neck, the pale greenish tail and wing coverts, but the breast and flanks are less distinctly spangled, the under parts being of a more uniform red.

The slight change in the plumage is doubtless owing to the influence of a change of climate acting through many generations, added, perhaps, by a change of diet. We are informed by Mr. J. English Torbett that the ripe seeds of the *Calla æthiopica*, so common as a greenhouse plant in this country, are much sought after by the pheasants in St. Helena, and that it forms a large portion of their food.

Closely allied to the ordinary Chinese pheasant is a bird which has been described as a distinct species by Consul Swinhoe, under the title of the Ringless Chinese Pheasant (*P. decollatus*). It was obtained by him at Chung-king-foo, in Szechuen, and a somewhat similar bird was procured by Père David, at Moupin, near the Thibetan boundary. I cannot regard these birds as anything more than mere local varieties of the ordinary Chinese species, and must refer those who wish to trace the slight distinctions between them to Mr. Elliot's " Phasianidæ," in which they are figured. In the same magnificent folio will be found engravings of the Mongolian Pheasant (*P. mongolicus*), the Yarkand Pheasant (*P. insignis*), and Shaw's Pheasant (*P. shawii*); all closely allied to the common Chinese species, if not merely to be regarded as geographical variations from it. None of these forms are known in a living state in Europe, and consequently do not require detailed notice in the present work.

CHAPTER XIII.

PHEASANTS ADAPTED TO THE COVERT (CONTINUED).

THE JAPANESE PHEASANT (*PHASIANUS VERSICOLOR*).

APAN, among the numerous objects of interest with which it has furnished Europe, has supplied us with the most gorgeous of the true pheasants—*P. versicolor*. It is doubtful, indeed, whether any of the gallinaceous group, magnificent as many of them are, can surpass this bird in resplendent brilliancy. The wonderful dark grass green of the breast, that no painter can equal, the dark blue of the neck, and the brilliant scarlet of the face, taken together, constitute one of the most effective combinations of colour to be found in the whole class of birds. This splendid addition to the fauna of Great Britain was utterly unknown in a living state in Europe sixty years since. In 1840 a few birds were brought to Amsterdam from Japan. Of these a pair passed into the possession of the Earl of Derby—the grandfather of the present Earl—a man whose memory as a zoologist will be green when party strife is forgotten. Of this pair the female died, and the breed was established by crossing the male with several females of the ordinary species, and then pairing the half-bred progeny with the old male, and continuing the breeding back until the offspring were no longer capable of being distinguished from the original bird.

At the death of the Earl the Knowsley collection came to the hammer. A number of the versicolor pheasants, including the original bird, were purchased by Prince Demidoff for his preserves in Italy, and others passed into the possession of Mr. J. J. Gurney, of Norwich, by whom they were introduced into the preserves of that country. Since that period other specimens have been imported, and at the present time the *P. vesicolor* is established as a denizen of many of our preserves.

In form, habits, and disposition the *P. versicolor* corresponds closely to our common pheasants. As a game bird it is, both in the covert and on the table, of undeniable excellence.

As the bird crosses freely both with the common and the Chinese species, it is desirable to give an accurate and detailed description of its plumage. For this purpose I shall again have recourse to Mr. Gould's "Birds of Asia," and reproduce his elaborate description of the two sexes:—

"The male has the forehead, crown, and occiput purplish oil green; ear tufts glossy green; chin, throat, and sides and back of the neck glossy changeable bluish green; back of the neck, breast, and under surface deep shining grass green, with shades of purple on the back of the neck and upper part of the breast; feathers of the back and scapularies chesnut, with buffy shafts and two narrow lines of buff running round each, about equidistant from each other and the margin; lower part of the back and upper tail coverts light glaucous grey; shoulders and wing coverts light greenish grey, washed with purple; primaries brown on the internal web, toothed with dull white at the base; outer web greyer and irregularly banded with dull white; tertiaries brown, freckled with grey, and margined first with greenish grey and then with reddish chesnut; centre of abdomen and thighs blackish brown; tail glaucous grey, slightly fringed with purplish, and with a series of black marks down the centre, opposite to each other at the base of the feathers, where they assume a band-like form; as they advance towards the tip they gradually become more and more irregular, until they are arranged alternately, and in the like manner gradually increase in size; on the lateral feathers these marks are much smaller, and on the outer ones are entirely wanting, those feathers being covered with freckles of brown; orbits crimson red, interspersed with minute tufts of black feathers; eyes, yellowish hazel; bill and feet horn colour.

"Compared with the female of the common pheasant, the hen of the present bird has all the markings much stronger, and is altogether of a darker colour. She has the whole of the upper surface very dark or blackish brown, each feather broadly edged with buff, passing in some of the feathers to a chesnut hue; those of the head, and particularly those of the back, with a small oval deep spot of deep glossy green close to the tip; primaries and secondaries light brown, irregularly barred with buff, and with buffy shafts; tertiaries dark brown, broadly edged with buff on their inner webs, and mottled with dull pale chesnut on the outer web, the edge of which is buff; tail dark brown, mottled with buff, and black on the edges, and crossed by narrow irregular bands of buff, bordered on either side with blotches of dark brown; on the lateral feathers the lighter edges nearly disappear, and the bands assume a more irregular form; throat buff; all the remainder of the under surface buff, with a large irregular arrowhead-shaped mark near the top of each feather; thigh similar, but with the dark mark nearly obsolete."

The habits of the Japanese pheasant in its native country were first described by Mr. Heine, the naturalist attached to the American expedition to Japan, and the following observations by him were published in Commodore Perry's "Japan Expedition":—"After the treaty of Yokuhama had been concluded, the United States squadron proceeded to Simoda. A friendly intercourse with the natives was established, and I constantly availed myself of Commodore Perry's kind permission to make additions to our collections in natural history. One morning, at dawn of day, I shouldered my gun and landed in search of specimens of birds, and that day had the good fortune to see, for the first time, the Versicolor pheasant. The province Idza, at the southern extremity of which the port of Simoda is situated, forms a long neck of land extending from the island of Niphon, in a southerly direction, and is throughout mountainous, some of the mountains being from 4000 to 5000 feet high. The valleys are highly cultivated, presenting in the spring a most luxurious landscape. The tops of the mountains and hills are in some places composed of barren rocks, and in others covered with grass and shrubs, producing an abundance of small berries. Between those higher regions and the fields below the slopes are covered with woods, having, for the greater part, such thick undergrowth that it is scarcely possible to penetrate them. Following the beautiful valley, at the outlet of which the town of Simoda stands, for about four miles, I came to a place where the Simoda creek divides into two branches. Selecting the eastern branch, I soon left fields and houses behind me, and ascending through a little gulley, I emerged from the woods into the barren region. It was yet early in the morning; clouds enveloped the peaks and tops of the hills; the fields and woods were silent, and the distant sound of the surf from the seashore far below rather increased than lessened the impression of deep solitude made upon me by the strange scenery around.

"The walk and ascent had fatigued me somewhat; I had laid down my gun and game-bag, and was just stopping to drink from a little spring that trickled from a rock, when, not ten yards from me, a large pheasant arose, with loud rustling noise, and before I had recovered my gun, he had disappeared over the brow of a hill. I felt somewhat ashamed for allowing myself thus to be taken so completely aback; but, noticing the direction in which he had gone, I proceeded more carefully in pursuit. A small stretch of table-land, which I soon reached, was covered with short grass and some little clusters of shrubs, with scattered fragments of rocks; and as I heard a note which I took to be the crowing of a cock pheasant, at a short distance, I availed myself of the excellent cover, and crawling cautiously on my hands and knees, I succeeded in approaching him within about fifteen yards. Having the advantage of the wind and a foggy atmosphere, and being moreover concealed by the rocks and shrubs, I could indulge in quietly observing him and his family. On a small sandy patch was an adult cock and three hens busy in taking their breakfast, which consisted of the berries already mentioned growing hereabouts in abundance. From time to time the lord of this little family stopped in his repast and crowed his shrill war-cry, which was answered by a rival on another hill at some distance. At other moments again, when the sun broke forth for a short time, all stretched themselves in the golden rays, and rolling in the sand, shook the morning dew from their fine plumage. It was a beautiful sight, and I looked upon it with exceeding pleasure; so much, indeed, that I could not find the heart to destroy this little scene of domestic happiness by a leaden shower from my fowling-piece. Suddenly the birds showed signs of uneasiness, and I soon discovered the cause in a Japanese root-digger coming from the opposite direction. I therefore took up my gun, and standing on my feet, raised the birds also, and as they flew towards the next hill, I had the good fortune to bring down the cock with one barrel of my gun, and one of the hens with the other.

"The Japanese, who came up after I had loaded my gun and secured my game, looked with some astonishment at the stranger, for I was certainly the first foreigner who had been in pursuit of game on the hunting grounds of Niphon. He evidently asked me several questions, which I was not, of course, able to understand, but from his signs, and the frequent repetition of the word "statzoo" (two), I inferred that he inquired whether I had fired twice in such quick succession with one gun. I nodded and explained to him as well as I could the nature of my double-barrelled gun, and the use of percussion caps, which seemed to astonish and delight him very much. A pipe of tobacco which I offered was gladly accepted; and in answer to a question that he appeared to understand, he gave me the name of the pheasant as Ki-zhi. Later in the day more people came to the hills, some for the purpose of digging roots, others to look after their cattle, which appeared to be turned out to graze on the hills. The birds had taken to the bushes, where I could not follow them, and so obtained no more specimens on that occasion.

"A few days after, Lieutenants Bent and Nicholson and myself made another shooting excursion to the hills, but although we saw many pheasants, but a single specimen was shot, and the birds appeared to be very shy. We observed several Japanese with matchlocks about the hills, firing away at a great rate. As we did not see either of them with game, and as the game-laws of Japan are very severe, so much, so, indeed, that their observance has been made a special article of the treaty with the United States, I concluded that the firing was only for the purpose of driving away the pheasants to places where they would be more secure from the strangers."

The three species of pheasants—the *P. versicolor, torquatus,* and *colchicus*—readily breed with each other, and the mixed progeny, from whatever parentage, are perfectly fertile. The effect of this introduction of foreign blood in our common breed has been amazing, producing an increase of size and vigour, and beautiful variations in the plumage, dependent on the species whose blood predominates in the cross.

Nothing can be more interesting than the production of these beautiful mongrels, which increase so rapidly that Gould stated his opinion that in twenty years' time it would be difficult to find a true species in this country. This, however, he regarded as of little moment, as fresh birds can always be obtained from their native countries, Asia Minor, China, and Japan. All naturalists, however, are not of Mr. Gould's opinion. The late Mr. Blyth informed me that *P. versicolor* and *P. torquatus* kept themselves distinct in two neighbouring copses at Lord Craven's, not intermixing, although at a comparatively short distance from each other, and that he believed, although these races will cross when in confinement, that in the open country the birds of each would select their proper mates and produce pure bred offspring, an opinion which I regard as exceedingly doubtful.

The cross between the Japanese and common pheasant is a bird of brilliant plumage, easy to rear, of greater size than the average of English birds, and the flesh is very tender and well flavoured. In Norfolk this very beautiful cross was introduced some few years back by Mr. J. H. Gurney, who bred most successfully, both at Easton and Northrepps, from the birds he obtained at the Knowsley sale and the common pheasant (though chiefly with the ring-necked cross), and produced magnificent specimens; and from the eggs being greatly sought after by other game preservers in his district, the race soon spread throughout the county. "From personal observation and inquiry, however," writes Mr. Stevenson, "during the last two or three years, it appears evidences of this cross, even in the coverts where these hybrids were most plentiful, are now scarcely perceptible; the strong characteristics of the Chinese bird apparently absorbing all the less marked though darker tints of the Japanese. One of these birds, killed in 1853, weighed upwards of four and a half pounds, and many examples, which were stuffed for the beauty of their plumage, will be found in the collections of our country gentlemen."

The absorption of the Japanese in the more common race is not surprising when the small interfusion of new blood is

168 PHEASANTS FOR COVERTS AND AVIARIES.

taken into consideration, but with the fresh introduction of new blood, and the care in the preservation of the cross-bred birds, there can be no doubt a permanent breed would result, bearing the same relation to the pure bred Japanese that the common ring-neck does to the pure blooded Chinese species.

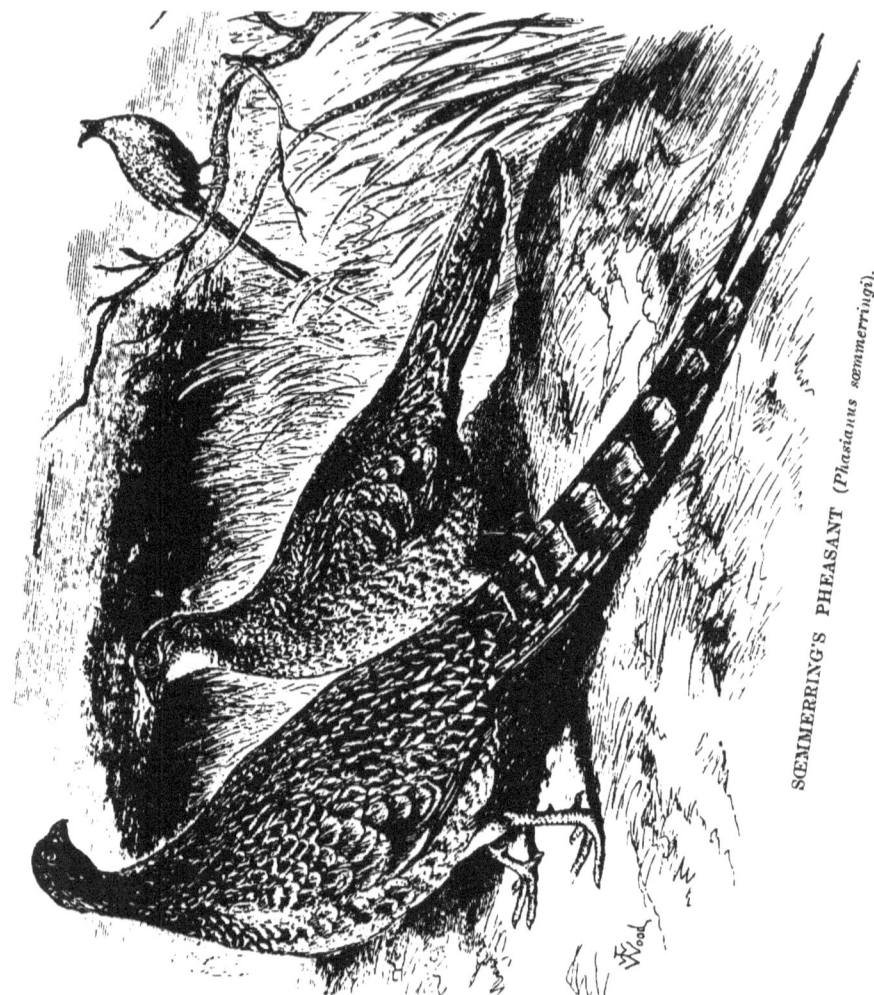

SŒMMERRING'S PHEASANT (*Phasianus sœmmerringi*).

CHAPTER XIV.

PHEASANTS ADAPTED TO THE COVERT (CONTINUED).

SŒMMERRING'S PHEASANT (*PHASIANUS SŒMMERRINGII*).

SŒMMERRING'S pheasant is a second exquisitely beautiful species inhabiting Japan. In the southern islands, Kin-Shin and Hondo, it is very numerous, and is commonly exposed for sale in the markets of Nagasaki. In other districts of the country its place seems to be supplied by the *Phasianus versicolor*. The bird was known to Temminck by the dried skins, but recently the living animal has been introduced into aviaries in Europe, and it has bred in the zoological gardens in London and Antwerp. In the Regent's Park Garden it first bred, according to Mr. Bartlett, in 1865, when the female laid ten eggs, but only a few birds were hatched, and the young birds died in a few days. Since then the breeding has been more successful, and mature specimens have been reared.

The species, however, is but ill-adapted to breed in confinement, as the males are excessively pugnacious—not only destroying one another, but even killing the females. This tendency is probably developed by captivity, and no doubt, if placed in a free range, Sœmmerring's pheasant would prove as fertile as the other species, but the experiment has never yet been tried. Mr. Bartlett, writing of this species in

Elliot's monograph, says: "Amongst the Phasianidæ some species are remarkable for their pugnacious and fierce dispositions; not only the males, but frequently the females destroy each other. The want of sufficient space and means of escape among bushes, shrubs, and trees is no doubt the cause of many females being killed when kept in confinement; and this serious misfortune is unhappily of no rare occurrence. After the cost and trouble of obtaining pairs of these beautiful birds, and they have recovered from their long confinement on the voyage, their owner is desirous of reaping a reward by obtaining an abundant supply of eggs as the birds approach the breeding season, when, alas! he finds that some disturbance has occurred, the place is filled with feathers, and the female bird, from which he expected so much, is found dead or dying, her head scalped, her eyes picked out, or some other serious injury afflicted. I have found some species more inclined to this cruel practice than others, the worst, according to my experience, being the *P. sœmmerringii.*" Mr. Elliot justly remarks that this is a sad account of such a beautiful bird, and he also suggests the right remedy when he states that doubtless this evil could be abolished by planting thick clumps of bushes in their inclosure, into which the hens could retreat and escape from the persecution of the males; if kept in large inclosures covered with shrubs, and filled with growing grass, there should be no difficulty in rearing these birds, especially if a due supply of fresh vegetable food be daily given.

Our knowledge of the habits of this magnificent bird in its native state is very limited. The best account which has been published is in Commodore Perry's "Japan Expedition" —one of those magnificent and expensive scientific works so liberally published by the American Government. Commodore Perry writes:

"This is undoubtedly the most beautiful of all the true pheasants, and will compare in richness and brilliancy of colour with almost any other species of bird. In the adult male the neck and back are of a deep golden red, with a metallic lustre of great beauty, but the female is exceedingly plain and unpretending.

"Like the Versicolor, the present is only known as a bird of Japan; and but few years have elapsed since it was first introduced to the attention of naturalists by the celebrated Professor Temminck, well known as the most distinguished of European ornithologists. It appears to inhabit the same districts of country as the Versicolor, and to subsist on much the same description of food; but we regret to say that the gentlemen of the expedition had no opportunity for observing this species to such an extent as to enable us to make any important contribution to its history.

"Nothing having previously been published in relation to this beautiful pheasant, we have exerted ourselves to obtain all available information, and have great pleasure in again acknowledging our obligations to Mr. Heine, the accomplished artist of the expedition, for the following note :

"' On one of my excursions I came very suddenly upon another species of pheasant, of very beautiful colours, and with a very long tail. Being in the midst of briars, and in an inconvenient position, I missed him, or at least did not injure him further than to shoot off his two long tail feathers.

"' Returning on board in the evening, I found that our chaplain, the Rev. George Jones, had purchased a pheasant of the same kind from a Japanese root-digger in the hills. It was not wounded, or otherwise injured, and seemed to have been either caught in a trap or found dead. To my inquiries of the Japanese Dutch interpreter, whether these birds were ever hunted, I could obtain but evasive answers; but if, however, such is the case, the right is undoubtedly reserved to the princes and nobility.

"' It appears that both these kinds of pheasants inhabit similar localities, and are abundant over the southern and the middle parts of the island of Nipon, for even during my rambles in the vicinity of Yokuhama, in the Bay of Yeddo

I could hear their calls in the little thickets and woods scattered over the country.'

"For the following note on the bird now before us, and the preceding species, we are indebted to the kindness of Joseph Wilson, jun., M.D., of the United States Navy, who was attached as surgeon to the squadron of the expedition :—

"'Our acquaintance with the pheasants of Japan began soon after our arrival at Simoda, or about the middle of April, 1854. A Japanese brought to the landing-place a young bird, which, with the dark tips on his downy covering, and his frequently repeated "peet-peet," might have been mistaken for a young turkey but for his diminutive size. This interesting little fellow had been obtained by hatching an egg of a wild pheasant, obtained in the hills, under a domestic fowl.

"'A few days after this a male pheasant in full plumage was brought to the same place, dead but uninjured, and evidently but very recently killed. The golden brilliancy of this bird's plumage is probably not exceeded by any object in nature, and is quite equal in lustre to the most brilliant markings of the humming-birds, or the most highly burnished metal. This splendid colouring covers the whole body of the bird, merely shaded with a little copper-red about the tips and margins of the feathers, so as to show the lance-head form of the feathers. This specimen was taken on board the flagship *Independence* and preserved.

"'The specimen of the other species that I saw was shot by Mr. Heine, who made a very beautiful painting of it. The two birds are found in the same localities, and seem to be similar in habits.

"'The Japanese system of agriculture, although very minute, and appropriating all available land to some useful purpose, yet affords abundant shelter for the native fauna. Scarcely any land is tilled except such as can be watered, so that the tops of hills and large portions of mountainous and precipitous places are appropriated to the growth of timber, or left covered with the primitive forest. These wooded districts afford shelter for wild hogs, foxes, and raccoons (the skins of which were seen), as well as for the pheasants; and they all descend in turn to plunder the crops, or steal the chickens in the valleys. During the first part of our stay at Simoda the cultivated fields afforded no food for the pheasants. The natives told us there were plenty in the hills; but no one was willing to undertake to show them, and several rambles through the bushes where these birds were supposed to feed ended in disappointment. Only once I had a glimpse of a brood of young ones near a hut in the mountains, but they immediately disappeared by running very rapidly. Perhaps one reason of our want of success is to be found in the fact that the wheat was ripe, and partially harvested before we left (June 24th), so that during the time of our efforts they were enabled to fill their crops occasionally from the wheat-fields, and lie very close in the hills during the day, without being under the necessity of wandering in search of food.

"'The note of one or the other of these species of pheasants was heard frequently. On the top of a precipitous hill, about a mile south of Simoda, covered by small pines and a very thick growth of shrubbery, a pheasant (so we were assured by the Japanese) passed the weary hours, while his mate was on her nest, and very sensibly solaced himself and her with such music as he was capable of making. It was, however, anything but melodious, and may be represented as a sort of compound of the filing of a saw and the screech of a peacock. There are two notes only, uttered in quick succession, and represented by the Japanese name of the bird—*Ki-ji*; but the second note is much longer, louder, and more discordant, in fact has more of the saw-filing character—*Kee-jaeae*. These two notes are uttered, and if the bird is not disturbed they are repeated in about five minutes. A good many attempts, perhaps twenty, to become better acquainted with this individual all failed. It seemed impossible to make him fly, though his covert was by no means extensive.'"

This species is readily distinguished by the widely separated transverse bands on the tail of the male, and the short, rounded tail of the female (8in. in length, that of the male being 23in.), the feathers of which are tipped with white at the extremity. We are not aware of any hybrids between this and allied species, although their production would be very interesting as bearing on a suggestion made by Darwin to the effect that "if the female Sœmmerring pheasant with her short tail were crossed with the male common pheasant, there could be no doubt that the male hybrid offspring would have a much longer tail than that of the pure offspring of the common pheasant. On the other hand, if the female common pheasant, with her tail nearly twice as long as that of the female Sœmmerring pheasant, were crossed with the male of the latter, the male hybrid offspring would have a much shorter tail than that of the pure offspring of Sœmmerring's pheasant."—"The Descent of Man," Vol. II., p. 156.

The following description of the two sexes is taken from Mr. Gould's magnificent folio, "The Birds of Asia":—" The male has the whole of the upper surface and throat of a fine coppery brown, with a lighter border to each feather, which in some lights appear of a purple hue; in others rich coppery red, and in others again bright but deep flame colour—this latter tint being especially conspicuous on the lower part of the back and upper tail coverts. This is the general appearance. On examining each feather singly, it is found to be grey at the base, dark rich brown in the middle, with a broad stripe down the centre, and on each side of dark coppery brown, with a lustrous stripe on each side of the tip; wing coverts the same, but devoid of the lustre at the tips; a few of the greater coverts with a narrow bar of creamy white at the tip, within which is a still narrower one of black. Primaries dark brown, crossed by irregular broken bands of a tawny hue; secondaries dark brown, freckled near the tip with tawny, and a large patch of deep rufous near the end of the outer web, becoming much paler at the extremity; on the tips at the inner webs of several of them the double mark of white and black, as on the greater coverts. Tail rich chesnut red with black shafts, and crossed at intervals of about two inches with a narrow irregular band of black, and a second broader and more decided band of the same colour—the space between the bands being of a similar but paler tint than the body of the feather; the second band of black, moreover, becomes broader, and gradually blends with the general colours of the feathers as they approach the extremity. On some the intermediate pale band is white; feathers of the under surface marked like the upper, but the bordering is not luminous, and terminates in dull grey, within which, on the lower part of the sides of the abdomen, is a narrow line of white; eye orbits red; bill brown colour; feet bluish-brown colour.

"The female has a patch of dark brown at the back of the head, with a narrow bordering of rufous at the end of each feather; feathers of the head and upper surface generally mottled with rufous, with a narrow edging of black at the tip, and with a stripe down the centre, which on the sides of the neck and shoulders is white, and on the other parts deep buff; rump and upper tail coverts deep rust red, each feather faintly barred with dark brown, some of the wing coverts marked at the tip with black and white, as in the male, but the marks are broader, and not so pure; throat deep buff, feathers of the under surface brown, largely striped down the centre, and tipped with pale or creamy buff, and bordered on each side with tawny; tail short, central feathers greyish brown, freckled with dark brown; lateral feathers rufous, crossed obliquely near the tip with dark brown, beyond which the end is white."

Under the title of *P. scintillans*, a variety of this pheasant has been described as a distinct species, but it appears to differ only in the male having the feathers on the back more or less completely margined or tipped with white.

Mr. W. R. Ogilvie-Grant, in his "Hand Book to the Game Birds," writes: "It can only be recognised as a well marked variety, for it not only occurs in the same islands where the *P. sœmmerringii* is found, but every intermediate stage of plumage between the two forms may be seen."

REEVES'S PHEASANTS (*Phasianus reevesii*).

CHAPTER XV.

PHEASANTS ADAPTED TO THE COVERT (CONTINUED).

REEVES'S PHEASANT (*PHASIANUS REEVESII*).

ARCO POLO, the old Venetian traveller, who returned to Venice in 1298, after a residence of seventeen years in Tartary, was evidently acquainted with the magnificent species now known as Reeves's Pheasant. In the language of his original translator, whose quaint orthography I have followed, he is made to state: "There be plenty of Feysants and very greate, for 1 of them is as big as 2 of ours, with tayles of eyght, 9 and tenne spannes long, from the Kingdom of Erguyl or Arguill, the W. side of Tartary." This description can only be applicable to the species now under consideration. From this time, until described by Latham and Temminck, this bird was comparatively unknown, except from the inspection of Chinese drawings. Sonnini, who preceded Temminck, concludes his account by stating that it is very possible that the bird, of which he had merely seen pictures, "exists only in the imagination of the Chinese painters." Singularly enough, the species was, for thirteen years—namely, 1808 to 1821—living in the aviary of Mr. Beale, at Macao.

Dr. Bennett, in his "Wanderings in New South Wales," states: "In Mr. Beale's splendid aviary and garden at Macao the beautiful *Phasianus veneratus* of Temminck, the *P. reevesii* of Gray, now commonly known by the name of the Reeves's Pheasant, was seen. It is the *Chee-kai* of the Chinese.

"The longest tail feathers of the bird are 6ft. in length, and are placed in the caps of the players when acting military characters. This I observed at Canton, where some of the beautiful tail feathers (rather in a dirty condition, like the actors themselves, who, in their tawdry dresses, reminded me of the chimney-sweepers in London on a May-day) were placed erect on each side of their caps as a decoration.

"The Chinese do not venerate this bird, as was first supposed, and which may have caused Temminck to bestow on it the name of *veneratus*; but it is superstitiously believed that the blood of the bird possesses poisonous properties, and that the Mandarins, when in expectation of losing their rank and being suddenly put to death by order of the Emperor, preserve some of it on a handkerchief in a dried state, on sucking which they fall down and instantly expire.

"Mr. Beale's first male specimen, obtained in 1808, was kept in a healthy state for thirteen years; after its death he endeavoured to procure others, but did not succeed until 1831, when four specimens were brought from the interior of China, and purchased by him for 130 dollars; these were, I believe, taken to England subsequently by Mr. Reeves."

The first living bird of this species was imported into Europe about the year 1831 by Mr. Reeves (of the firm of Dent and Co.). This specimen was a male. The son of this gentleman, Mr. John R. Reeves, brought a female over in 1838, and the pair were in the Zoological Gardens at same time; but the male being old, they did not breed. Some cross-bred birds were reared from the hen, who died in 1840, these are now in the British Museum.

Dr. Latham, in his "General History of Birds," gave a description of this species from a drawing and tail feathers in the possession of Sir J. Anstruther. He states:—"I had an opportunity of seeing a bundle of thirty or forty of these tail feathers, which were brought from China, and I found amongst them specimens of every length from 18in. to 7ft." The species was named by Latham *P. superbus*. Temminck described it under the title of *Faisan superbe* in his "Pigeons et Gallinacés," published in 1813. At this date it was known to him only by the two central tail feathers, and the drawings of native Chinese artists. Subsequently, however, he obtained a skin of the male, which he figured in his "Planches coloriées," giving it the erroneous name of *P. veneratus*. This plate was copied on a reduced scale in Jardine's " Naturalist's Library," published in 1834. Dr. J. E. Gray, in his " Indian Zoology," named the bird after the gentleman by whom it was introduced into England, and by this name it is now generally known.

The successful introduction of the living birds now in England is owing to the combined efforts of the late Mr. John J. Stone and Mr. Walter H. Medhurst, H.M. Consul at Hankow. Owing to their exertions, this splendid pheasant is now firmly established in this country, and like the *P. versicolor* and *P. torquatus*, is to be seen at large in our woods, and specimens are not unfrequently to be bought in the wholesale markets.

For several years Mr. Stone made continuous efforts to obtain this and other new pheasants from Northern China, but with no satisfactory results until the aid of Mr. Medhurst was obtained. It is mainly due to that gentleman's thorough knowledge of the natives of China, and of their language, that the true habitat of this bird was ascertained, and an experienced Chinaman sent into the interior for the purpose of collecting this and other rare pheasants, of which coloured drawings had been supplied for his guidance.

The first three lots of birds obtained all died before reaching England, with the exception of one male, which lived for about three months. The fourth lot was obtained in the direction of Syechney, about thirty days' journey from Hankow, and from it seven Reeves's pheasants were deposited in the Zoological Gardens, Regent's Park. Mr. Medhurst was anxious that Her Majesty the Queen should have early possession of specimens of *Phasianus reevesii*; and, in compliance with his wish, one male and two females were offered to and graciously accepted by Her Majesty. Since the successful reintroduction of these birds they have bred freely both in confinement and at large in England and on the Continent, and are now to be purchased at the dealers.

With regard to the distribution of this bird in China, Mr. Saurin remarks:—"The Reeves's pheasant, called by the Chinese *Chi-Chi*, is very rarely seen in the Pekin market. For a long time I failed to discover from what quarter they came. . . . . Last winter I ascertained, however, that they came from the Tung-lin; and I have reason to suppose that they are to be found nowhere else in the province of Chi-li. About twenty birds were brought down alive last winter. They are never brought in frozen or by Mongols. Their flesh is very delicious, and superior, to my taste, to that of any other pheasant."

The general character of the plumage of the Reeves's pheasant is well shown in the illustration. The head is covered by a cowl of white, surrounded by a band of black, with a spot of white under the eye; the neck has a broad ring of white; the feathers of the back and upper part of the breast are of a brilliant golden yellow, margined with black; those of the lower part of the breast are white, each one presenting bands of black more or less irregular in their arrangement; the under parts of the body are deep black; the tail is formed of eighteen feathers, which are closely folded together, so that the entire tail appears narrow; at the broadest part the feathers are about 2in. in breadth; the ground colour of each tail feather is greyish-white in the centre, and golden red at the edges, and crossed with crescent-shaped bars, which vary in number according to the length of the feather, in the longest feathers being considerably more than fifty.

A very interesting observation was made by the late Mr. Blyth on the voice of this species. He states :—" I have heard the call-note of Reeves's pheasant, and it was some time before I could satisfy myself that it actually proceeded from such a bird. It is like the simple song of some small passerine bird, delivered in as high a key as the song of the hedge sparrow (*Accentor modularis*), one of which happened to be singing at the same time. A repetition of the same note seven or eight times over, quite musical but not loud, being as unlike what would be expected from such a bird as a pheasant, as the voices of sundry *Columbidæ* are utterly different from what would have been expected to proceed from pigeons and doves."

The late Mr. J. J. Stone, to whom naturalists are so much indebted for his introduction of this and other splendid pheasants, was of opinion that the value of Reeves's pheasant in this country rests mainly upon its size and strength of flight, making it the prince of game birds for our woods. In a communication to me on the subject, he wrote : " The point I aim at is to induce the large landed proprietors and game preservers to introduce the Reeves's pheasant into their coverts, believing that it will (from its wild character) afford the best sport of all the pheasants, and from its size and the magnificence of its plumage it must be a desirable addition to our list of game birds. I want to see Reeves's pheasant common on the dinner table; and there is no reason why it should not be so in a few years, seeing that it is now being bred freely in Belgium, and may be purchased there at about the price which the Versicolor still commands, though much longer introduced into Europe." Since Mr. Stone's death several successful attempts have been made to introduce this most noble of all the true pheasants into our coverts.

Lord Lilford, writing in March, 1881, gave me the following information : " I have kept several in pens, and found them very uncertain layers, although one season my hen birds laid an average of twenty eggs apiece, from which a very good proportion of young birds were hatched out. My impression is that these birds lay best—at all events in captivity—at about their third or fourth year. My gamekeeper, who has had charge of them, assures me that the young birds are very hardy and easy to rear. I have in Northamptonshire (the county in which I have tried these birds) no very large extent of woodlands of my own, and cannot therefore tell you much of their habits in a wild state, as they are very much given to roaming to great distances, and a good many have fallen victims in my neighbours' woods, besides the large percentage that may be always allowed in a foxhunting country. They have certainly crossed, though not abundantly, with the common pheasant. The male hybrid of the first cross is a most splendid bird. Reeves's pheasant is a very wild, shy bird, very quick on the wing, somewhat given to go back if possible, but quickly attains a good height in the air, giving good rocketing shots. I found them most excellent for the table—in my opinion far superior to the common pheasant. I believe, from what I have seen and heard of this species, that for real success with them in this country a wide range of hill coverts would be most eligible. I believe that Sir Dudley Coutts Marjoribanks has had great success with Reeves's pheasants in Inverness-shire." I am informed that as many as sixty Reeves's have been shot in these coverts in a single season.

Fifteen years after Lord Lilford favoured me with the above communication he published in "The Birds of Northamptonshire" a further report on this species, in which he maintains its desirable character as a game bird for ranges of woodland in mountainous districts. His account is as follows:

"Another most beautiful species, known as Reeves's or the bar-tailed pheasant (*Phasianus reevesii*), though we have found it hardy, easy to rear, and excellent for the table, our opinion is that, as it possesses the roaming instinct in a still higher degree than the ring-necked species, and is of a very

REEVES'S PHEASANT (*Phasianus reevesi*)

wild and wary nature, it is not a desirable bird from a sporting point of view, except in very large ranges of woodland; and from what we have heard from a friend, who has been in the native haunts of this fine bird in the mountains of Northern China, we are inclined to think that it is more likely to prosper in Scotland and in Wales than in our own Midlands. The hybrids between this and the common pheasant are beautiful birds, but not, so far as we have been able to ascertain, prolific."

The late Mr. Horne, of Hereford, who reared numbers of the Reeves's pheasants, forwarded to me a letter from a lady who has been most successful with them in the extreme north of England regarding this species. This lady writes:

"The cock and two hens I purchased have done wonders, and my estate is now fairly stocked with birds, having put all this season's eggs in nests of the common pheasant, except a few which I reared myself and a few which I sold. My hens last season averaged nearly fifty eggs each—not bad laying."

Not only in the extreme north, but in the more cultivated parts of England, Reeves's pheasants have done well. One gentleman informs me that during the year 1895 he raised more than twenty in the open, which are now all in full plumage, and that he found them easy to rear.

There can be no doubt whatever, as suggested by Lord Lilford, that, the bird being from North China, is hardy and well adapted to mountainous districts, such as those of Scotland and Wales. It appears that the easiest way of introducing it as a wild bird in those places to which it is adapted would be to place the eggs in the nests of pheasants breeding in the open. Reared under those circumstances, the young would be hardy and vigorous in the extreme, and would be much more likely to do well than if hand-reared and turned out afterwards. The fact of the hybrids between it and the common species being sterile is, to my mind, rather in its favour than otherwise. There would be no mongrel crosses introduced, and Reeves's pheasant could be confined to those regions to which by its size and habits it is specially adapted. With regard to its beauty and magnificence there can be no doubt, and Lord Lilford speaks practically as to its value as a bird for the table, but I have never had the opportunity of testing its value in this respect.

The most important communication respecting the value of the Reeves's pheasant as a game bird, and its rearing in the forests of mountainous districts, was made to the *Field* on February 9, 1896, by Mr. J. G. Millais. This was accompanied by a most graphic sketch of the flight of the bird, which Mr. Millais has kindly given me permission to reproduce. Mr. Millais's letter is as follows :

"I noticed a letter by Mr. Tegetmeier in the *Field* of January 25, on the desirability of establishing Reeves's pheasant as a British game bird; and as I have seen and shot several of these birds at home, perhaps my observations on the species may be of some interest.

"There is no game bird, I think, in the world, which, if introduced into suitable localities, would give greater pleasure to both the sportsman and the naturalist than this grand pheasant; for grand he certainly is, both to the eye as well as the object of aim to the expectant shooter. We all know, when a cock Reeves's pheasant attains his full beauty and length of tail, what a splendid bird he is as he struts about in his gorgeous trappings, and shows himself off for the benefit of his lady-love, but when the same bird is launched in the air, and dashes along above the highest trees of a wild Scotch landscape, leaving poor old Colchicus to scurry at what seems but a slow pace behind him, I can assure your readers that both the dignity and the pace are alike wonderful, and a sight not easily to be forgotten.

"Until the year 1890 I had seen and shot several Reeves's pheasants, and under ordinary conditions of covert shooting was content to consider the bird hardly a success from a gunner's point of view. During that autumn, however, I went to the annual covert shoot at Guisachan, Lord Tweed-

REEVES'S PHEASANT IN FLIGHT.

mouth's beautiful seat, near Beauly, in Ross-shire, and it was there, amidst the wildest and shaggiest of Scotch scenery—in country which must to a great extent resemble the true home of the bird in question—that I had cause to alter my opinion.

"In one high wood of old Scotch firs, on a steep and broken hillside above the waterfall, the sight of these birds coming along only just within gunshot, in company with common pheasants and blackcocks, I shall never forget. I say, 'in company with,' but, as a matter of fact, as soon as one of the long-tailed sky-rockets cleared the trees, he left the others far behind, and came forward at a pace which was little short of terrific. I doubt if any bird of the genus goes faster.

"Now, this is all that the sportsman wants. Here we have a bird of unrivalled beauty, great hardihood, and unequalled pace, which practically fulfils all the conditions which the modern shooter requires. The only other condition which is absolutely essential to make the bird a success from this point of view is its local environment. In this respect Guisachan is not singular, and I could name a hundred localities in Scotland, England, and Wales where Reeves's pheasant would be certain to succeed.

"The Guisachan birds were obtained by the late Lord Tweedmouth from Balmacaan, the late Lord Seafield's estate near Loch Ness, where I have also seen them shot. No artificial rearing was resorted to; the birds were breeding in a wild state, and shifting entirely for themselves, except for the maize which was put down for the ordinary pheasants. At Balmacaan, where the birds were in low open woods, one may see Reeves's pheasants killed in the way in which they should not be. Here these birds (as is the case when turned down on any ordinary English preserve) have formed most undesirable habits. It is with great difficulty they can be got to rise at all, and when this is effected they keep low, and afford no sport whatever. Now, at Guisachan all this is obviated by the rough nature of the ground. There is heavy bracken, fallen trees, mountain burns, and, above all, rough heather. These cause the birds to get up almost at once. The trees being high and dense assist their elevation, and force them to a respectable height from the very start.

"In conclusion, I should like to make one observation on the flight of Reeves's pheasant which I have never seen touched on before, and which is both interesting and remarkable. Reeves's pheasant has the power to stop suddenly when travelling at its full speed, which may be estimated at nearly double that of an ordinary pheasant; and this is performed by an extraordinary movement when the bird makes up its mind to alight on some high tree that has taken its fancy. This bird may be said to be furnished with a 'Westinghouse brake' in the shape of its tail, otherwise the feat would be impossible. By a sudden and complete turn of the body, both the expanded wings and tail are presented as a resistance to the air, and the position of the bird is reversed. This acts as an immediate buffer and brake, and by this means the bird is enabled to drop head downwards into the tree within the short space of eight or ten yards. This is such a very remarkable movement, and one which of necessity requires some illustrative explanation, that I send you herewith a sketch of it, which may be of interest."

Mr. J. Mayes, head-keeper to the late Maharajah Dhuleep Sing, writing from Elvedon, in 1877, stated: "I have bred the Reeves's pheasant for the last five or six years, rearing them by hand, and have had pretty good luck with them the last two years, having succeeded in rearing about sixty in the two seasons; but I find they are much healthier turned out than when penned up. The soil here is dry and sandy, which seems to suit them very well. Two years ago I penned up fifty very fine young birds, about half-grown; but they swelled very much about the head, and went completely blind, and about twenty of them died, but those that we have turned out seem to be in very good health and condition. As regards hybridizing, I know they will do so, as three years ago a hen

Reeves escaped from the pens, bred with a common pheasant, and brought up five very fine young birds, much larger than the common pheasant, and of beautiful plumage."

Many specimens of hybrid or cross-bred Reeves have been reared in confinement. That figured in the same plate with the Bohemian pheasant was the offspring of a male Reeves with a Bohemian hen; it partook, as may be noticed, the characters of both species, the tail being of intermediate length, the white cowl, cheek patch, and neck ring of the Reeves being retained, but the splendid golden yellow of the body being almost entirely wanting.

Hybrids have been produced between a male Reeves's pheasant and female Cheer (*Phasianus wallichii*), but they have little beyond their size to recommend them. In appearance they look like dirty faded Reeves's, with comparatively short tails. They are of large size, like the parent species, and would in all probability partake of those terrestrial habits of the Cheer which preclude its being advantageously introduced as a game bird, as it often refuses to rise, even when hunted or pursued with dogs.

A singular hybrid was produced and described by Mr. R. Sanders, of Heavitree, who writes: "I have three most beautiful male birds, bred between the Reeves and gold. The size is about that of the male Reeves, but the plumage does not in the least partake of that of either parent; it is very much like that of the copper pheasant of China (of which I had several some years since), but not so dark. The chief colour is a soft light brown, running into a light copper; the marking on the head is somewhat after the Reeves; the tail very long."

CHAPTER XVI.

PHEASANTS ADAPTED TO THE AVIARY.

THE GOLDEN PHEASANT (*THAUMALEA PICTA*).

AMONGST the birds that are reared in our aviaries on account of the beauty of their plumage, the two species of the genus *Thaumalea* occupy a very prominent position. These birds have been separated from the more typical pheasants (which have been already described as constituting the restricted genus *Phasianus*) by several well-marked characters, the most conspicuous of which are the presence of a crest of silky feathers on the crown of the head, and a tippet of broad flat feathers encircling the upper part of the neck. The Golden Pheasant (*Thaumalea picta*) has been long known in captivity in Europe; it was described by Linnæus under the name of *Phasianus pictus* in 1766, but of its habits in its native country nothing whatever had been ascertained; even its exact locality was doubtful until more recent explorations in China. It is now known to inhabit the mountains of the western central districts, and it has been shot by Europeans on the banks of the Yang-tsze, one hundred miles north of Hankow. In the north of China it is, according to Père David, quite unknown.

In its mature plumage the male is one of the most gorgeous of the whole tribe. The head is ornamented with a long crest

of silky orange-coloured feathers. This extends backwards over a tippet formed of broad flat feathers, which are of a deep orange colour, with dark blue bars across the tips; these latter form, when the feathers are in position, a series of horizontal lines across the tippet. During the courtship of the female this collar or tippet is brought over to the side nearest the hen, as shown in the background of the engraving of this species; the late Mr. T. W. Wood paid more attention to the amatory displays of birds than any other writer. Respecting that of the Golden Pheasant he writes: "Not the least remarkable example of the lateral mode of display during courtship is that of the Golden Pheasant, whose elegant form and brilliant colouring are so well known in this country. The male runs very playfully after the female, and placing himself in front of her, quickly expands his collar, bringing nearly the whole of it round to the side where it is to be exhibited, and thereby presenting to view a flat disc of bright orange-red, banded with perfect regularity by blue-back semicircles; the hen on seeing this frequently runs away pursued by her would-be mate, who generally finds himself placed with his other side towards her, and the collar is accordingly shown on that side. At the moment the full expansion of the collar takes place, the bird utters a very snake-like hiss, which, according to our notions, would not be very fascinating as a love-song; the body is very much distorted, as is the case with the true pheasants, but the tail is not spread so much, as the curved, roof-like shape prevents its forming a flat surface. Slight breaks would occur in the black stripes of the collar when expanded, were it not that each feather has a second black stripe which is so placed as effectually to prevent this."

Below the tippet on the lower part of the neck the feathers are deep-green margined with velvet black; below this again are the scapular feathers of a dark crimson; the back and rump are golden yellow; the tail itself is very long, the two longest central feathers are covered with small irregular circles of light-brown on a dark ground, giving them a mottled appearance; the other feathers are barred diagonally with dark brown on a lighter ground. On each side of the base of the tail extend the long narrow upper tail coverts of a bright orange crimson. The wings when closed show the deep blue tertiaries covering the chesnut secondary quills. The upper part of the throat is light-brown; the breast and under parts orange-scarlet. Taken altogether, its appearance is so remarkable that it looks more like one of the bizarre creations of Chinese fancy than a real bird. The birds of this genus differ from the true pheasants, in the fact that the mature masculine plumage is not assumed until the autumn of the second year; the young cocks looking, during the first twelve months of their lives, very much like the hens, from which, however, they can be readily distinguished by pulling one or two of the feathers of the neck, which are reproduced of the distinctive masculine character.

The hens are very plain and unobstrusive, being barred with alternate shades of light and dark brown. When barren, they, like the other birds of the family, assume the more gorgeous apparel of the male.

Under the name of the Black-Throated or Java Golden Pheasant (*Thaumalea obscura*) a variety of this bird has been described as "a good species." It has never been obtained in a wild state and is evidently merely a variety that, like the black-winged peacock, may appear at any time amongst birds of the ordinary type, and could never be regarded as a species by those who have studied the subject of variation practically. It differs merely in the upper part of the throat being darker in colour and obscurely spangled, in the pattern of the mottling of the upper tail feathers, and in the general darker hue of the females and young.

One of the best and most complete accounts of the habits and management of the Golden Pheasant in confinement is that written by Mr. W. Sinclaire, of Belfast, and published in Thompson's "Natural History of Ireland." Mr. Sinclaire writes:—

"Golden Pheasants are very easily reared in confinement, and are quite as hardy as any of the other pheasants, or as any of our domestic fowls; indeed, I question if any of them are sooner able to provide a subsistence for themselves, or to live independent of the parent bird. In the several years' experience I have had in the rearing of these birds, I have considered them past all danger when they arrived at the age of three or four weeks; in fact, at that age those which I brought up in the garden began to leave the bantam hen which hatched them, and take into the gooseberry bushes to perch at night; and very soon after into the apple trees. I always observed that they roosted at the extremity of the branches, where they were quite safe from the attacks of cats or other vermin. This habit, together with their very early disposition to roost at night, leads me to infer that their introduction into this country as a game bird would not be difficult; and that in our large demesnes, where protected from shooters, they would become very numerous. But I should imagine that they would not answer where the common pheasants were already introduced, as they are shy, timid birds, and would be easily driven off by the other species. The individuals before referred to, which were reared in the garden, consisted of a family of six; they always remained in the garden, where they were regularly fed, except at the commencement of winter, when they ceased roosting in the apple trees, took to a belt of Scotch firs which bounded the garden on one side, and roosted in them all the winter and following spring. I have seen them sitting in the trees when the branches were laden with snow, but they did not seem to suffer in the slightest degree from the severity of winter. About the month of February they first began to wander from the garden for short distances; and as the spring advanced, finally disappeared, and I never could hear of their being met with afterwards.

"In rearing the young I found that the very best food for them, and of which they were most fond, was the larvæ of the bluebottle fly, with a quantity of which I always was prepared prior to the young being hatched. I took care to have a constant supply during the season by hanging a cow's liver over a barrel, in the bottom of which was some bran or sawdust, into which the maggots dropped. A fresh liver was hung up about once a week. In addition to these larvæ, the young were supplied with potatoes, alum curd*, groats, and Indian corn meal; this last I found they were very fond of, and it seemed to agree with them particularly well. It was mixed into the form of soft dough with a little water, which was all that was required. They were also constantly supplied with green food, such as lettuce, when they were in the aviary. But the best way is to have a coop, railed in front, into which they are put with the hen twenty-four hours after they are hatched. This coop should be placed upon a gravel walk as near to the windows of the house as possible, so that they may always be within observation; a small verdure garden is the best possible locality, as the young have plenty of range, with shelter under the bushes from both sun and rain. In the instance which I have already alluded to, the hen was allowed to range about six feet from the coop, by means of a small cord attached to a leather strap round one of her legs, and the other end tied to the coop; the young pheasants never wandered far from the hen, and always came into the coop to remain with her at night. In front of each coop a small frame was put down, boxed round on three sides, without a bottom, and railed at top; the open side was put close to the coop, and the young birds could run through the rails of the coop into the inclosed space, and were safe from the night attacks of cats, rats, &c. This frame was always kept before the coops for the first few days after the young

---

* Custard prepared as described at page 111 will be found far superior to curd.

were hatched, and until they became acquainted with the call of the hen. When I first began to rear young pheasants I could not at all account for their seemingly foolish manner for the first two or three days after being hatched; they would run gaping about without appearing to notice the hen or her calls to them to come for food. The reason of this I afterwards believed to have been owing to their ignorance of the language of their foster-mother, which it took some time for them to understand; during this process it is necessary to keep them confined within the frame before their coops, as, were they to wander a few yards from the hen, they would not heed her call, and would inevitably perish.

When three or four weeks old, it is necessary, if reared for the aviary, to pinion them, which is done by cutting off rather more than the first joint of the wing, having previously, by means of a needle and thread, inserted close to the small wing-bone, and brought round the large one, just within the skin, taking up the main blood-vessels; the piece of the wing is then chopped off on a block. There is no loss of blood, and I never could observe that the birds seemed to suffer in the slightest degree afterwards, although the operation I daresay was painful enough. My reason for taking off rather more than the first joint of the wing was because I found that if only the first joint was taken off, the birds were always able, when grown up, to get out of the aviary, which was about 12ft. high; and I found it thus requisite to take off so much as to render them incapable of any attempt at flying, but I left enough remaining to enable them to reach their roosting-place at night. I furnished them with a kind of ladder by nailing cross pieces of wood on a long piece about 3in. wide, and which they very soon learned to walk up and down with facility. One aviary in which I kept some had a back wall to it covered with old ivy, and they preferred roosting in this; indeed, I always found that, although during a wet day those which were at liberty took shelter under a roof, yet at night they would not do so, but would instead roost in the open air. The females will lay about twenty-five eggs each in the aviary. I always provided them with baskets to lay in, which they only sometimes made use of; they take twenty-four days to hatch. The young cocks do not attain their full plumage until after the moult of the second summer; they drop their chicken feathers when about three months old; their plumage is then something like the hen's, but sufficiently bright in some parts as easily to distinguish them from the young females. In general there are more cocks than hens.

"If the cock birds are placed in a portion of the aviary apart from hens, any number may be kept together. I have had so many as twelve males in full plumage together, and when during the summer (and indeed at all times) these beautiful birds were going through the very curious and fanciful attitudes and manœuvres peculiar to them, it was one of the most brilliant sights to be observed in nature. The flashing of their various golden, crimson, blue, and purple plumes in different lights was absolutely dazzling to the eye, and at these times they contrive to display all the most beautiful parts of their plumage to the utmost advantage; the golden crest is raised; the splendid orange and purple-tipped collar is spread out to its full extent, while the scarlet tail coverts are shown in all their beauty. During the whole time the birds are leaping and dancing round each other, and uttering occasionally their peculiar shrill cry."

Mr. Thompson states that he has never known the Golden Pheasant to live longer than ten or eleven years, and that such as came under his knowledge "died almost instantaneously, and when in the highest possible condition as to flesh and plumage," death being apparently induced by high condition and over fatness.

Respecting their management in aviaries still more confined for space, my friend the late Mr. Ed. Hewitt kindly gave me the following notes :—

"As I kept Golden Pheasants many years with success, a few

hints may, to beginners, be acceptable. They may with proper care be reared perfectly tame, but have always a tendency to be alarmed at the sudden appearance of strange dogs, cats, or even individuals; for which reason I think it advisable they should be pinioned if allowed an out-of-door run, lest they may be tempted to fly away, or on to the surrounding buildings; for, unlike common poultry, they are only tempted to return with great difficulty, as the moment they get from their accustomed range they seem as wild and uncontrollable as birds reared in a state of nature. Pinioning can be easily managed without the slightest detriment to the appearance of the bird. Let it, too, be always kept in mind, in handling pheasants never to lay hold of the legs or wing, for injury is certain to ensue; but take up the bird with both hands tightly round the body over the wings. This is the only safe way of capture, and they then may be taken about without injury at all, either to their plumage or to themselves.

"I would advise aviaries for their accommodation to be covered in entirely from the rain, as nothing tends so much to keep them in perfect feather; and then it will not be by any means difficult to guard them against another great annoyance—that of cats prowling about during the night and at twilight. From this cause numbers of pheasants of either kind have been destroyed, not from an actual hurt received from the cat, but from the birds in their fright flying furiously against the roof or the wirework, and scalping themselves. This may be prevented by letting a 'tar-sheet' be fixed closely every night, to cover the whole of the open work of the aviary. It has this double service: it prevents sudden rain wetting the sanded floor and causing damp (producing rheumatism in the inmates), and by being opaque prevents the shadow of passing cats being seen; for if they see cats at night the birds will fly, and thus seriously damage themselves. I found simple canvas for this purpose of no use whatever, being semi-transparent; the tar-sheet is effective from its density. It is on moonlight nights that the greatest danger is to be feared, for on these occasions the cats come very long distances, attracted no doubt by scent, and when they have once found your birds will be sure to pay them almost nightly visits. As the birds are valued for their beauty, it will add considerably to the perfection of their plumage to place a sufficiency of perches for their accommodation; not spare and thin ones, but made of deal spars about 1½in. square, the sharp edges being taken off with a plane. This will prevent their tails rubbing, and, whether intended for attraction or sale, add not a little to their value.

"In selecting the brood stock, a cock with four or even five hens will be a fair proportion. I always prefer a cock bird of the second year and hens of the same age (because they lay far more eggs), though the eggs of pullets of the preceding year are productive. The young hens will only lay ten or twelve eggs in a season, but the older birds when carefully managed will frequently lay thirty to forty eggs in the same period. These eggs require a longer incubation than those of common fowls, as they generally hatch on the twenty-fourth day, though I have repeatedly known them continue in the shell a day longer; therefore, if desirous of rearing a chicken or two with them (to insure greater familiarity), the fowls' eggs must be deposited accordingly, as nothing tends so sadly to unsettle a hen at hatching time as some portion of her chicks coming a day or two previously to the remainder, and it not unfrequently leads to the desertion of her nest.

"The eggs laid in an aviary should be at once removed from Golden Pheasants directly they are laid; the cocks being especially inclined to peck and eat them the moment they are produced. The best remedy I know is to procure half a dozen artificial eggs, and let them lie about always, and then the birds, seeing them constantly, regard them less. They are raised in confinement much more easily than the common pheasant, the young growing with great rapidity if well and frequently fed on custard, boiled eggs, good cheese—all chopped fine—and mixed with bruised hemp and canary seed. The maggots produced in flesh from the blow-fly will tend very greatly to their rapid growth. I am perfectly aware that ants' eggs are preferable, but when these are not available maggots will be found an excellent substitute, and should be given daily till the poults are somewhat grown. Wheat, hemp, and barley are the best food for the old stock. It is somewhat singular that neither variety will agree comfortably with the common pheasants in a wood; notwithstanding, I have seen the hybrids produced between these birds and the common pheasants. They are very beautiful, being of a strikingly rich auburn, shading into every variety of gold colour; but they were invariably unprolific, and sought every possible opportunity to evince their pugnacity to all other birds confined with them.

"Golden Pheasants will endure every severity of our climate. Some years since I gave away some eggs, from which birds were hatched and turned loose in a large plantation; they bred freely the ensuing year, and well stocked the preserve; the year following some withdrew to a covert at some considerable distance, driving away the common pheasants and taking possession of the whole. Some Golden Pheasants' eggs, which I forwarded as a present to a friend whose preserves are among the largest in the kingdom, were hatched very early last season and turned loose; these bore all the rigours of winter as well as any others, but in the spring began to show a decided aversion to their fellows of more sombre hue."

Mr. Hewitt further stated that the sexes in the chicks were easily distinguished, the eyes of the cocks being light, those of the hens deep hazel.

Golden Pheasants that have escaped to the coverts and been shot, are found when cooked to be of very delicate flavour. These escaped birds will sometimes breed with the ordinary species. Mr. Mayes, writing to me from Elvedon, stated that "the Golden Pheasant will cross with the common pheasant; we have had two or three instances of their doing so the last two years; but it is rare, as we have had Golden Pheasants turned out the last ten years or more, and never knew them to cross with any other pheasant before."

AMHERST PHEASANT (*Thaumalea amherstia*).

CHAPTER XVII.

PHEASANTS ADAPTED TO THE AVIARY (CONTINUED).

THE AMHERST PHEASANT (*THAUMALEA AMHERSTIÆ*).

LADY AMHERST'S Pheasant was first made known to Europeans by two male specimens presented by the King of Ava to Sir Archibald Campbell, and by him given to Lady Amherst, who retained them in India for about two years, and succeeded in bringing both alive to England, where, however, they lived only a few weeks. These specimens were figured and described under the title of *Phasianus amherstiæ* by Mr. B. Leadbeater in the "Linnæan Transactions" for 1828. Since that time until recently no living specimens have been seen in Europe, and in 1863 the male was figured in Mr. P. L. Sclater's list of *desiderata* required by the Zoological Society. The successful re-introduction of this remarkable species is entirely owing to the combined efforts of Mr. J. J. Stone and Mr. W. Medhurst, Her Majesty's Consul at Shanghai, who obtained twenty specimens in Western Yunan, eight of which reached Shanghai alive, and six—five males and one female—were successfully located in the Zoological Gardens, Regent's Park, in July, 1869. Since that time other specimens have been obtained, and the species has bred freely in confinement, and even in the open covert.

The general appearance of the species is strikingly beautiful. The accompanying engraving, though giving very correctly the general character, necessarily fails in imparting any idea of the coloration of the male. The irides are light, the naked skin of the face is light blue, the feathers of the forehead are green, but the long plumes which form the crest are crimson. The tippet, which is so characteristic a feature in the bird, is white, each feather being margined with a dark green band, and having a second narrow band at some distance from the tip. The front of the neck, the breast, shoulders, back, and wing-coverts are of an exquisite metallic green, each feather being tipped with velvety black. The lower part of the breast and belly are white, the thighs and under tail coverts mottled dark brown and white. The feathers of the rump have the exposed parts bright saffron yellow. The tail coverts are brown at the base, striped green and white in the middle, and brilliant scarlet at the ends. The two upper middle tail feathers have a light ground marked so as to resemble lace, with broad transverse bands of green about an inch apart. The other tail feathers have the inner webs mottled black and white, the outer webs with curved green bars, about three-quarters of an inch apart. The bill is pale greenish, and the feet and legs bluish lead colour. The female closely resembles the hen of the last species (*T. picta*), being a rich chesnut brown, with bars of dark brown, which are broader than those of the Golden Pheasant hen, and the under parts are lighter in colour; moreover, the bare skin of the face is pale blue like that of the male, but much smaller. The size of this species is somewhat larger than that of its close ally, the Golden Pheasant. In the male the adult plumage is not assumed until the autumn of the second year.

When Mr. Gould gave his description of this pheasant in his "Birds of Asia," the male only was known, and he wrote :—" It would give me great pleasure to see a female of this fine bird, and every ornithologist would be truly gratified by the arrival of any information respecting the part of the Celestial Empire in which it dwells, and any details as to its habits. The bird would doubtless be as easily kept in our aviaries as its near ally, the Golden Pheasant; and it is my ardent wish to see it thus located before I leave this lower world for the higher and brighter one which is the end of our hopes and desires." Every ornithologist must feel glad that Mr. Gould had his wish gratified.

Since the arrival of Mr. Stone's specimens, Mr. Anderson, the curator of the Indian Museum at Calcutta, has received skins of both sexes from Yunan and Upper Burmah, where it is not rare, the plumes being worn by the natives.

The only account of the habits of this beautiful species in a wild state occurs in a letter from Monsieur Carreau, a French missionary in Thibet, to the Paris Acclimatization Society. He states :—"The pheasant *Houa-ze-Ky*, the Flower Pheasant of the Chinese, always inhabits very rocky places. Whenever I have seen this bird flying upwards, I have always been able to shoot it; but if it was descending, I could not procure it, for then it disappeared with excessive rapidity. After having pursued it several times, I have found it more convenient to obtain it in the same manner as the natives, who lay in wait for it during the winter and catch it in snares. When the mountains are covered with snow, and the streams frozen, the Flower Pheasants are obliged to descend to the plains for water, but as soon as they are satisfied they ascend again. In the paths these birds follow each other in a line; and as they go in flocks, and the snares are few in number, the Chinese do not make much from the plumage and flesh of this beautiful pheasant. Ta-lin-pin is situated in the 29th degree of latitude N., and the 102nd degree of longitude E. : the heat of these places is very great, as they are surrounded by high mountains, and with very little vegetation. The mountains are covered with brambles, briars, and thorns, and also with grassy places; in these spots the Amherst Pheasant is met with in abundance. It is an error to think that, like other pheasants, it is met with in the forests; I have never found it there, and as in the neighbourhood of Ta-lin-pin it only exists where there are no forests, I doubt very much if bushy tracts are to its liking. The more rocky and desolate the mountains, the more certain are you to find the Flower Pheasants, in companies composed of from twenty to thirty individuals.

"The habits and economy of the Amherst Pheasant naturally accord with the places in which it delights; it is an extremely wild bird. Last year I kept one of these pheasants in a stable covered with straw; it hid itself so frequently and so well that once I was more than fifteen days in the belief that it was dead. I fed it with bread and rice, and it became very fat. If this bird should be introduced into Europe, it would be useless to endeavour to make it comfortable, if it has not in the aviary some place where, at the least noise, it can hide itself, otherwise I doubt if it can be preserved. I think, from the temperature of the mountains it inhabits, that the climate of France would be suitable for the Flower Pheasant. These particulars respecting the Lady Amherst's Pheasants are perfectly exact, since I have myself frequently hunted, captured, fed, and raised them. They would increase easily in Europe, provided they were not too much exposed to the heat of the sun, and that shrubs were grown in the aviary to allow their hiding when frightened."

The breeding of the Amherst Pheasant offers no difficulty, provided it be attempted under natural conditions, and not in the close pens, and stifling, vermin-haunted hatching-houses that are characteristic of some of our zoological collections. Not only has the pure race been increased, but the males have also bred freely with the hens of the Gold Pheasant (*Thaumalea picta*), and produced hybrids which are of surpassing beauty. At the sale of the surplus stock in the Zoological Gardens at Antwerp in 1872, a single male hybrid of this kind, in full plumage, realised 35*l*. The cross-bred specimens combine in a remarkable degree the most attractive features of the two species from which they are derived, and are unquestionably far more beautiful than either; compared with them the pure bred Amherst looks pallid, and the Gold Pheasant wants the beautiful contrast of the white neck tippet and the brilliancy of the green and blue.

The crest is fully developed, being larger than in either parent species; in colour it is a brilliant scarlet orange. The neck tippet is white, margined with brilliant dark green, resembling that of the Amherst, but considerably more developed. The iris, which is white in the latter species, is of a pale straw colour in the hybrids, as is the naked skin under the eye. The neck under the tippet, as well as the throat, is a resplendent green. The breast, which in the Amherst is white, is a brilliant scarlet orange, with a narrow transverse band of lighter yellow about an inch below the margin of the green feathers of the throat. The flanks are of the same colour as the breast. The back is yellow, running into the bright scarlet orange of the tail coverts and side sickle feathers. The wing coverts are of a magnificent dark steel blue. In all the characters mentioned, the hybrids possess the most gorgeous hues of the two species conjoined. The tail, however, is an exception; that of the Amherst is certainly more beautiful than that of the Gold, which latter, however, appears almost unchanged in the cross-breeds, but of somewhat increased size. As, however, in the so-called species *Thaumalea obscura*, the tail of the Gold tends to vary towards the markings of that of the Amherst, and the upper part of the throat to assume a spangled character, there would be no difficulty in breeding this cross with the Amherst tail. The cross-breeds are remarkably tame, feeding readily out of the hand.

Mr. Elliot, in his monograph of the *Phasianidæ*, gives a life-sized coloured plate of this hybrid, and acknowledges that "in size and brilliancy of dress he eclipses" both the parent species, adding: "Contrary to my intention of not figuring any hybrid pheasants, I have been induced to show this one, merely from its great beauty and the comparative rarity of at least one of its parents; but at the same time I cannot but believe that all those who breed pheasānts, either for pleasure or profit, would best consult their own interests by keeping their birds as pure in blood as possible, allowing no foreign strain to intermingle, and resolutely setting their faces against even such a magnificent impostor as here offers himself for our admiration." I quote this passage as illustrative of the beauty of the birds, although I differ entirely from the conclusions arrived at by the writer. There can be no possible doubt of the perfect fertility of the half-bred Amhersts. Mated with the pure Amherst, three-quarter pure-bred birds are the result; these show very little trace of the Golden species. The half-bred Gold and Amherst are equally fertile when mated with one another, and an intermediate breed may be perpetuated, which possesses the united beauties of both parent species, and be perfectly permanent in its characters.

The perpetuation of permanent races produced by the union of two perfectly distinct species is well known to all who do not wilfully shut their eyes to those facts which do not square with their theories. The late Mr. E. Blyth, a most accurate observer, and whose knowledge of species was unsurpassed, informed me that over a large extent of India no other domestic goose is known except the cross between the Chinese species, the *Anser cygnoides*, and the domesticated variety of the grey-lag, *Anser ferus*.

In the case of the true pheasants, *Phasianus colchicus*, *P. torquatus*, and *P. versicolor*, every variety of interbreeding takes place, and the intermediate forms can be perpetuated as may be desired; or, as was originally the case with the *P. versicolor* in this country, the pure breed can be established from a single individual.

Most naturalists maintain that these three pheasants are perfectly good species; but what is the test of a species?

For my own part, I am sufficiently heterodox in my belief to regard all the true restricted pheasants, such as *P. colchicus, versicolor, torquatus, shawii, mongolicus, elegans*, &c., as mere geographical variations of one type, capable of breeding together and perpetuating any cross that it may please experimenters to produce; and in the same manner the two species of the genus *Thaumalea*, namely, the Gold and Amherst pheasants, may be regarded as geographical races capable of yielding a permanent race intermediate between the two.

These views, which I maintained at the time of the publication of the first edition of this work in 1873, have been fully borne out by later experience. In March, 1881, Mr. A. D. Bartlett, the superintendent of the Zoological Gardens, wrote to me: "The hybrid Amherst and Gold pheasants breed freely *inter se;* but, as far as I can learn, in most cases the breeders have been breeding the half-bred hens with the pure Amherst males, for the purpose of obtaining as near as possible the characters of the pure Amherst; and this is very quickly accomplished, for in the third generation all traces of the Gold pheasant are lost, or nearly so."

The late Mr. Horne, writing to me in 1879, said: " With regard to the Gold and Amherst pheasants being turned out by landed proprietors, I know of a place in Ireland where there are large numbers of these birds breeding together in a wild state, and some of the crosses are very beautiful. There is also an estate in Scotland where Amhersts have been at liberty for years, and the owner wrote me they kept their own quarters, not allowing the other pheasants to interfere with them. I agree with you that it is a pity more of these birds are not turned out, as they form a great attraction to pleasure grounds. The easiest way to bring it about is to place a brood or two under hens in the kitchen garden; and, as they become fully grown, they naturally fly to the adjoining trees, and continue to hang about the place afterwards."

CHAPTER XVIII.

PHEASANTS ADAPTED TO THE AVIARY (CONTINUED).

THE SILVER PHEASANT (*EUPLOCAMUS NYCTHE-MERUS*) AND ALLIED SPECIES.

UNDER the name of *Euplocamus nycthemerus* the Silver Pheasant has been known to naturalists since the time of Linnæus. In the earlier works on natural history, such as that of Albin, published in 1738, and Edwards, in 1751, it was termed the Black and White Chinese pheasant, which name was employed by Buffon; it was also termed the Pencilled and Lineated Pheasant, and by Temminck, the *Faisan bicolor*.

Its native locality was first definitely ascertained by Consul Swinhoe, who informs us that it inhabits the wooded hills in the interior of southern China. Writing to Mr. Elliot, he states: " This bird is known to the Chinese as the *Pih Heen*, and it is one of those which are embroidered upon the heart-and-back badges of the official dresses of the civil Mandarins to denote the rank of the wearers. So far as I have ascertained, it is found in the wooded mountains of the following provinces: Fokein, Canton, Kwangse, and Kweichou. It is brought to Canton city from the province of Kwangse by the west river, and offered alive in the shops for sale. All the birds I have seen so offered have been captured; I do not think the Chinese had the bird in confinement. A friend of mine shot one in some woods, in the

mountains about 100 miles from Amoy (Province Fokein), but I have never met with the species in my rambles."

In his recent handbook on Game Birds, Mr. W. Ogilvie-Grant writes: "According to the Abbé David the Silver Pheasant is becoming very rare in a wild state, and is only found in South China, towards the North of Fokien, and perhaps in Chekiang. He says that most of the Golden and Silver pheasants that one sees at Shanghai come from Japan, where these two Chinese species are reared in captivity. The Silver Pheasant is known in China by the names of Ing-ky (Silver Fowl) and Paé-ky (White Fowl). Very little indeed is known of the habits of this extremely fine species in a wild state, though it has long been one of the commonest aviary birds. The males are, unfortunately, so extremely pugnacious and such big heavy birds that they fight with, and often kill, any other male pheasant living in the same aviary, and for this reason must be kept separate."

From their large size, commanding appearance, and the beauty of the markings, Silver Pheasants have long been favourites in our aviaries. They have the additional recommendation of being exceedingly hardy, of laying freely in captivity, and of being easy to rear when young. They also become perfectly tame, feeding freely from the hand. These birds could be readily domesticated, if it were thought desirable to do so. I have known several instances where they have been allowed to run at full liberty, and have seen the birds sufficiently tame to come and stand before a window, waiting for their accustomed treat at the hands of the members of the family. The hens, though not usually allowed to do so, will readily hatch their own eggs, and attend upon their chickens with all the care of common fowls. I have recently seen a pair, belonging to Mr. Clarence Bartlett, in a moderate-sized aviary, the hen of which had laid, hatched, and was rearing a strong healthy brood of young, the cock being active in defence of his family, and attacking most viciously any person going into the inclosure.

No game cock could be more determined or courageous in his behaviour: and the sharp spurs with which this species is armed render his assault a thing to be avoided, as he would fly at the face of the intruder on his domain.

From the readiness with which the Silver Pheasants can be domesticated and reared among the other denizens of the poultry yard, they occasionally escape into the coverts and become wild, under which conditions they breed freely. They are not, however, desirable additions, either to our limited stock of game birds, or, though exceedingly ornamental, to our very restricted number of domestic poultry, inasmuch as they are exceedingly pugnacious, driving away all the true pheasants from the preserves, fighting with the fowls, killing the young chickens in the poultry yard, and not even hesitating to attack dogs, children, and even grown-up persons during the breeding season. When wild they are flushed with difficulty, and on the wing they have been rightly characterised as being most unsatisfactory from a sporting point of view, flying dangerously low, in a horizontal direction but a few feet from the ground.

With regard to their edible qualities I can speak very positively, as I have had specimens that have been shot in the coverts cooked as pheasants, and found them destitute of the flavour of game, and altogether of very inferior quality. The flesh was white, and, although the bird had been well hung, exceedingly firm.

A correspondent informs me that he has "reared several Silver Pheasants in confinement, and has turned them out about the grounds. The males are exceedingly tame, but also exceedingly dangerous. Last year I had a lovely specimen, which used to feed at the window of the breakfast-room with the peafowl and other birds, and even knock at the glass and make its way into the room. But in the spring, when hatching was going on, he attacked ladies and children in the most determined manner, always flying at the face. He would dodge people walking, and make his appearance from under the bushes in a very unexpected manner. On one occasion he knocked a lady down, and on another occasion entered the drawing-room and attacked a lady who was sitting there."

Another writer says :—" I have for many years had a score of them running loose with the poultry—two cocks, one an old one, the other a young one of last year, just getting into full plumage ; the others are hens. In bad weather and in winter they roost in the poultry house, at other times in the trees. The males are most pugnacious and jealous, fighting and bullying the fowls—so much so that I am obliged to have their spurs cut off—and the hens very spiteful to young poultry. The others I have shut up, otherwise they would fight until they killed each other. In the breeding time they are shut up in large pens.

" I have frequently had the hens sit on and hatch their eggs; when they have young ones, if anyone goes near them they act like partridges. I have seen them charge dogs and drive them away. I have also seen a cock watching a fox stalking him, and when the fox made his rush the bird flew over him, but lost his tail. To show how severely they can make these spurs tell, one of my keepers kicked at an old Silver cock pheasant to drive him away, when the bird turned on him and sent his spur right through his boot. They are quite as bad as peafowls in a kitchen garden ; they will eat all the fruit. They are not very good birds for the table, but they are useful as being eatable in February and March."

The Silver Pheasant is a long-lived bird, even in confinement. Mr. Thompson, in his "Natural History of Ireland," states that he has known one live twenty-one or twenty-two years in captivity.

The male, without possessing the gorgeous coloration of many of the *Phasianidæ*, is a very beautiful bird. The face is entirely covered with a bright vermillion skin, which during the spring becomes excessively brilliant, and is greatly increased in size, so as to almost resemble the comb and wattles of a cock; the flowing crest is blue-black, the bill light green.

The upper part of the body is white, pencilled with the most delicate tracery of black. The whole of the under parts are bluish-black, the legs and feet red, the spurs well-developed and usually very sharp. The female is smaller than the male; her general colour is brown, mottled with a darker tint; the crest and tail are much less ample than those of the cock; the outer tail feathers are light, marked with black on the outer webs. The female in confinement usually lays from eight to fourteen eggs, and the young are most easily reared under a common fowl.

The genus *Euplocamus*, to which the Silver Pheasant belongs, includes several species. They are distinguished from the true pheasants by the crest, by the more fowl-like form of the tail, and by the males (and sometimes even the females) being strongly and sharply spurred. The common species, the kaleege or kalij of India, breed very freely, even in confinement, but are not adapted for turning into the covert, as they rise with difficulty, and their flesh is not equal for culinary purposes to that of the ordinary pheasant. A correspondent writes:—"I have been shooting lately in preserves where, amongst other game, I had the pleasure of seeing the kaleege on the wing. The birds had been bred under hens from eggs taken from old birds in a mew, treated in the same manner as pheasants, and were at this time—the last week in December—practically as wild as the pheasants in the same coverts. A more unsporting-looking bird on the wing I never met with, or a more unsatisfactory one to knock down. Its flight is low, never rising more than eight or ten feet from the ground, and therefore in a line with everybody's head, consequently a most dangerous bird in a *battue*. Its flight is more like that of a coot or moorhen than any bird I know; the slow, noiseless flight, and the dark plumage, making it very like the former bird. It runs much before rising—is very savage, driving away the other game birds, and is the most unsatisfactory game bird I ever saw. My friend with whom I was shooting is therefore killing them down."

SPECIES OF KALEEGE.   211

Twelve different species of kaleege have at various times been shown in the Zoological Gardens, Regent's Park. Of these the greater number have bred either with their own species, or have produced hybrids with other *Euplocami*. Amongst those that breed the most freely may be mentioned Swinhoe's pheasant (*E. swinhoii*), the purple kaleege (*E. horsfeldi*), the black-backed kaleege (*E. melanotus*), and the white-crested (*E. albo-cristatus*). The different species of *Euplocami* hybridise together even in a wild state, and there is no difficulty in rearing a very large series of hybrids in captivity.

CHAPTER XIX.

PHEASANTS ADAPTED TO THE AVIARY (CONTINUED).

THE EARED PHEASANT (*CROSSOPTILON MANTCHURICUM*).

F the remarkable group of birds known as the Eared Pheasants, constituting the genus *Crossoptilon*, five species are known, though only two, the Mantchurian (*C. mantchuricum*) and the white Tibet species (*C. tibetanum*) have been received in Europe in a living state.

The Eared Pheasants differ in many very essential particulars from the more common species. Both sexes are alike in plumage, and are only to be distinguished by the presence of spurs on the legs of the males. The large size and peculiar character of the tail coverts separate them from any allied group. The first specimens seen alive were presented to the Zoological Society by Mr. Dudley E. Saurin, in 1866; since that time others have been imported, and a considerable number have been bred in this country and on the continent.

The Mantchurian Eared Pheasant is more remarkable for the singular arrangement of its plumage than for brilliancy of colouring, in this latter respect not approaching the gorgeous hues of the true pheasants, or many of the closely-allied birds. The general colour of the body is a sombre brown; the true tail feathers are white, with dark tips; but the bird derives its remarkable appearance from its large size

and the peculiar character of the tail coverts, which spring from the lower part of the back, and in great part obscure the true tail. These tail coverts are white, and have the barbs separated, so that they form an elegant appendage to the body. The legs and feet of the Eared Pheasant are red in colour, and of true scratching or rasorial type, the claws being bluntly curved, like those of the common fowl. The head is very striking in its general appearance; the vaulted beak is of a pale fleshy white, contrasting strongly with the red skin of the face, which again is thrown into prominence by the white feathers that constitute the so-called ears of the bird.

Consul Swinhoe states that, "This bird is found in the hills north of Pekin, in Mantchuria, and brought in winter to Pekin in large numbers, both alive and dead. It is called by the natives the Ho-ke. The feathers of this bird were formerly worn by Tartar warriors. I have not seen the species in its wild state."

Père David informs us that these birds frequent the woods of high mountains, and that they subsist much more upon green vegetables, leaves of trees, and succulent roots than upon grain. In their habits they are more gregarious than the common pheasants, assembling together in flocks of considerable size. In domestication they become exceedingly tame, feeding readily from the hand. When at large they appear remarkably hardy; they breed when only one year old, and acquire their adult plumage at the first autumnal moult.

They possess the very rare instinct of domestication. 1 have seen specimens at Mr. Stone's residence in the Welsh hills as familiar as barn-door fowls. In the closely confined pens in our Zoological Gardens their increase has not been very rapid, but they have proved themselves as hardy and prolific as common turkeys would have been if placed under similar disadvantageous circumstances. Mr. Bartlett writes: " Of the Crossoptilon we have reared nine fine birds the second hatch, having lost by the gapes the first brood of seven."

By placing a young brood in a large walled-in garden, where they could obtain abundance of fresh vegetables and insect food, they should offer no more difficulty in rearing than barn-door fowls; all they would require would be custard and lettuce in addition to ants' eggs, if obtainable; but fed on dry hard corn, and kept in small aviaries with brick floors, success is not to be expected.

Of the allied species, Hodgson's Crossoptilon (*C. tibetanum*) three specimens were living in the Zoological Gardens in 1891. In this the general colour is bluish-white, but the crown of the head is black, the wings dark, and the tail black crossed with green and blue. It is a native of Tibet.

Under the name of *C. drouynii*, a species very closely allied if, indeed, it be not identical with the last, has been described and named by M. Verraux, and figured in Elliot's Phasianidæ.

The original Eared Pheasant described by Pallas was a slaty-blue species. Pallas's specimens have long been lost, but recently, owing to the indefatigable exertions of Père David, skins have been received at the Museum at Paris, and the original *C. auritum* is now known to be perfectly distinct from the Mantchurian species, with which we are most familiar in the living state.

THE MONAUL (*Lophophorus impeyanus*).

CHAPTER XX.

PHEASANTS ADAPTED TO THE AVIARY (CONTINUED).

THE IMPEYAN PHEASANT (*LOPHOPHORUS IMPEYANUS*).

HE Monaul, or Impeyan Pheasant, is one of the most gorgeous birds; the wonderful metallic brilliance of its plumage, "gleaming in purple and gold," never fails to attract the attention of the spectator. In the Zoological Gardens it has bred frequently, but a native of the Himálas, seldom descending far below the snow line, and suffering from the heat of summer, is not likely to succeed on the London clay. During the life of my friend, Mr. J. J. Stone, I saw at large on the Welsh Hills Impeyan Pheasants as tame as the other poultry, and I have little doubt but that in suitable localities, as in the North of Scotland, this magnificent bird might be introduced to advantage either as a domestic or wild bird.

Should it be thought desirable to try the experiment in any appropriate locality, this can only be done by a consideration of their habits in a wild state, and I have therefore great pleasure in quoting the following from the late Col. Tickell, who was well acquainted with the birds in their natural haunts:—

"The Monaul ranges high in the mountains where it is found, keeping near the line of snow; and although met with in the ridges next the plains, becomes much more numerous farther in the mountains. It frequents the entire range of the Himála, from Afghanistan to Sikhim. Its range in elevation varies according to season, but in the severest winter it does not appear to descend below 6000 feet above sea level. I have seen numbers in Nepal in winter, brought with other kinds of pheasants by the Botias for sale in the plains of India, where they soon perish when the hot weather begins.

"They are forest birds, and difficult to be found in summer when vegetation is profuse, unless by ascending to the highest limits of the forest, when shots may be obtained in the open downs above, and amongst the rocks and thin herbage near the snow. In autumn, as the underwood decays, they descend and scatter through the woods, sometimes in great numbers, and seek lower levels as the winter advances and the soil becomes frozen. At such times they draw near to the small villages, perched on the lower spurs and above the sheltered valleys, and seek their food in the fields, where the mountaineers, with their large hoes, have dug up the soil. In these seasonal migrations it has been remarked that the females and young birds descend lowest and approach nearest to human habitations.

"They appear to be either capricious in their rambles through the woods, or are actuated to particular spots at particular times for reasons not apparent. Sometimes the sportsman will put up in one part of the forest fifteen or twenty in the space of four or five acres. In another portion he may keep on flushing for the rest of the day single birds, feeding in solitude, far apart. At no time are they gregarious, and whenever alarmed they rise and escape independently of each other. In some parts only cock birds are found, in others only hens.

"Severity of cold and scarceness of food have their taming effect on the Monaul, as on other birds, and the lower the snow the easier the task of making a bag. When on the wing, it generally flies a long way, and if much alarmed crosses over to a parallel ridge. Occasionally, however, it will settle on the low limb of a tree, at no great distance, and once there, it is, like many other gallinaceous birds, easy of access.

"Sometimes when approached in open spots it walks off, or begins to run, stopping often and eyeing the intruder, till suddenly, and without apparent immediate cause, it will rise with a startling flapping or flutter of the wings, scattering the dead leaves in a shower around, and fly headlong into the wood with a succession of short, piercing, shrieking whistles, which appear to act as a warning to some distant companions, for their calls are often heard in reply. When feeding quietly and in security the Monaul has a sweet mellow call—a long plaintive note—which it utters from time to time, especially of a morning and after sunset. It has the same melancholy effect on the ear as the creaking whistle of the curlew winging his way along the mudflats as evening settles over the lonely shore. The call has a rather melancholy sound, or it may be that as the shades of a dreary winter's evening begin to close on the snow-covered hills around, the cold and cheerless aspect of nature, with which it seems quite in unison, makes it appear so.

"The Monaul breeds towards the end of spring. The courtship is carried on in the chesnut and large timber forests before the birds ascend, during the summer heats, towards the regions of perpetual snow. It is generally near the upper limits of these forests, where the trees are dwarfed and sparingly scattered, that the hen lays and incubates three to five eggs, in a depression on the ground. The eggs are of a dull cream or pale buff colour, sprinkled with reddish brown. Like most gallinaceous birds, the Monaul may be said to be omnivorous. Those I have had in confinement ate rice and grain readily, as well as insects, worms, maggots, flesh, lizards, fish, eggs, &c. It is a diligent digger, and the slightly expanded tip of the mandible acts like a hoe or shovel. I had several of these birds in an aviary at Mullye, in Tirhoot. They were strong and vigorous as long as the cold weather lasted, and soon became tame, and did not succumb to the atmosphere of the plains till June, when the rains had set in. Unlike the smaller hill pheasants, they were not pugnacious. If shipped off early in the cold weather from Calcutta, these birds could easily enough be transported to England, where the temperature would suit them, if there were any means of giving them shelter during the extreme severity of winter, or of procuring for them in that season a proper substitute for the insect food which never fails them on the lower elevations of the Himála. If they could become as thoroughly acclimated as the common pheasant, they would indeed be a superb ornament to our parks and plantations, though perhaps no great acquisition to the table. It is many years ago since I tasted the Monaul, and, speaking from memory, the flavour appeared to me much the same as that of peafowl; the breast being tender and palatable in the young birds, but no part being fit for anything but soup in old specimens. The Monaul has bred in England, both in the Zoological Gardens of London and in the possession of the Earl of Derby, where the female is said to have laid on one occasion thirteen to fourteen eggs."

In appropriate localities there should be little difficulty in rearing the young, which should be amply supplied with custard and ants' eggs, in preference to much grain, and the fowl rearing them should be allowed as much freedom as possible, in order that she may supply the young chicks with appropriate insect food.

The following is the description of the two sexes and young:—" The bill of the male is dusky brown or horny; iris sombre brown; legs greenish lead colour; naked orbits; small blue head; crest and throat green, and highly metallic; the lanceolate feathers on the hind neck amethystine or bright purple, changing in lights into cupreous green with a golden glance; middle of the back white; but all the rest of the upper parts, including the upper tail coverts, rich blue, glancing with green and purple, highly glossed, the purple predominating on the back and rump, the green on the wing and tail coverts; remiges plain black; tail pale rust colour; all under parts black, and without gloss. The female is entirely cinnamon brown; the feathers shafted pale, and irregularly barred and marked sepia; primaries blackish; chin and throat white. Entire length of the male, about 24 inches; wing, 11; tail, 7¼. The female is a little smaller. The young males are at first like the female, but may be distinguished by the black spots on the chin and throat. They assume the adult plumage gradually, and in irregular patches scattered over the body."

Mr. W. Ogilvie-Grant, in his hand-book of the Game Birds, asserts that Gould was in error in calling the common Monaul the Impeyan Pheasant, *L. impeyanus*, a name which should be applied to another species—the Chamba Monaul. It is to be regretted that the name under which one species has been so long known should be transferred to another in scientific catalogues. There is no doubt whatever that under the name of Impeyan Pheasant the Monaul will long be recognised, as little or nothing is known of the Chamba species, the female being entirely unknown. In addition to these there are two other Monauls, that named after De L'huys from West China, and another named after Dr. Sclater.

CHAPTER XXI.

PHEASANTS ADAPTED TO THE AVIARY (CONTINUED).

THE ARGUS PHEASANT (*ARGUS GIGANTEUS*).

HE Argus Pheasant, as it was termed by Linnæus, is undoubtedly one of the most magnificent of the family of the pheasants. Its native haunts are the forests of Malacca and Siam, and it is also found in North-western Borneo. It is so extremely shy in its habits that it is rarely, if ever, shot, even by native hunters, who nevertheless manage to secure numbers by snaring the birds.

Mr. Wallace, in his most interesting work on the Malay Archipelago, describes his journey into the heart of the Argus country, and, writing of Mount Ophir, fifty miles eastward of Malacca, states :—

" The place where we first encamped, at the foot of the mountain, being very gloomy, we chose another in a kind of swamp, near a stream overgrown with zingiberaceous plants, in which a clearing was easily made. Here our men built two little huts without sides, that would just shelter us from the rain, and we lived in them for a week, shooting and insect-hunting, and roaming about the forest at the foot of the mountain. This was the country of the great Argus Pheasant, and we continually heard its cry. On asking the old Malay to try and shoot one for me, he told me that,

though he had been twenty years shooting birds in these
forests, he had never yet shot one, and had never seen one
except after it had been caught. The bird is so exceedingly
shy and wary, and runs along the ground in the densest parts
of the forest so quickly, that it is impossible to get near it;
and its sober colours and rich eye-like spots, which are so
ornamental when seen in a museum, must harmonise well
with the dead leaves among which it dwells, and render it
very inconspicuous. All the specimens sold in Malacca are
caught in snares, and my informant, though he had shot none,
had snared plenty."

The great peculiarity of the birds of this genus is that the
secondary flight feathers of the wings are excessively enlarged and lengthened, being in the males double the length
of the primaries, and covered on the outer webs with the
singular ocellated spots from whence the bird derives its
name. In the male, also, the two central tail feathers are
extremely elongated, and project in a very singular manner
beyond the others.

Until recently the *Argus giganteus* was the only known
species in the genus; but another smaller Argus (*A. grayi*)
is now known by specimens in the British Museum; and the
existence of one or two others is suspected from specimens
of feathers, differing from those of the known species.

The great Argus is over five feet in length, the tail being
three feet eight inches long. The prevailing colour of the
plumage is ochreous red or brown, unrelieved by any lively
or brilliant shade. The tints are distributed with so much
harmony, and covered with such a profusion of small spots,
or even points, sometimes darker and sometimes lighter than
the ground, that they produce the most agreeable effect.
Its long and broad secondary feathers are covered in their
entire length by a row of large eye-like spots, closely
imitating half globes; the colour of these, as that of the
plumage, has, however, something resembling ancient bronze.
The primary feathers, with whitish external barbs, speckled

with brown, and with inner barbs of the colour of a fallow deer, dotted with white, have their shafts of the most beautiful sky blue. The naked skin of the face and neck is bright blue, and contrasts well with the bronze hue of the plumage. The female neither exhibits the extraordinary development of the tail and wings nor the eye-like spots of the male. Her plumage is darker, and the total length is only twenty-six inches.

The two specimens (a male and female) figured in the engraving had been living some few years in the Zoological Gardens in the Regent's Park when the first edition of this work was published, at which time only five specimens of the Argus had been seen alive in Europe; since then it has been more frequently imported, and a dozen adult specimens have been received in the Zoological Gardens, and several young have been bred there. In addition to those in the Regent's Park, others have lived in the possession of the King of Italy, and in the Zoological Gardens at Amsterdam. It is singular that the Argus, although so exceedingly shy when wild, becomes perfectly tame in captivity, returning to its aviary when allowed to escape, as related by Lieut. Kilham in the *Ibis* for 1881.

The ornamentation of the secondary wing feathers in the male Argus is one of the most wonderful in the whole animal kingdom; the ornamental marks are usually termed ocelli or eyes, but they much more closely resemble ball and socket ornaments. As these ocelli are not visible when the wing is closed, the mode in which they were displayed has hitherto rather been conjectured than described, and even in recent works the bird has been portrayed displaying its plumage in a perfectly unnatural manner.

Fortunately, however, the pair of Argus pheasants formerly in the Zoological Gardens, Regent's Park, were closely watched for some days in succession by the late Mr. T. W. Wood, who had several opportunities of seeing the male bird display the magnificence of its plumage, and made a drawing of it at the time. At my request he kindly favoured me with the following particulars:

"It is with great pleasure that I comply with your request to give you a description of the mode of display of the Argus. The male bird commences by running about very briskly, bending his neck, and seeming to look at the female 'out of the corner of his eye;' he is evidently at this time in a very playful mood: he elevates his wings (while still closed) and shakes them. Suddenly, when close to the female, he throws his wings forward, the primaries resting on the ground, the secondaries extending upwards, and the tertials having their upper surfaces pressed together. At this time slight rustling sounds are heard, which I have no doubt are produced chiefly by the movements of the side feathers of the tail, as they are alternately moved outwards and inwards; the large feathers of the wings are also slightly waved, and moved at regular intervals downwards towards the female. But the most remarkable circumstance is that the bird places his head behind, or under one wing, so that in front there is nothing to intercept the view of the observer of his plumage. With the head so placed, how is he to observe his 'ladye love,' which, one would think, he must very strongly desire to do? My idea was that, by lowering his head a little, he could peep between his wings; but Mr. A. D. Bartlett has told me that he has seen the head thrust through the wing feathers, and Mr. E. Bartlett suspected this on finding some secondary feathers of a specimen which he set up disordered at their bases. I have drawn the head in the position in which it has been placed when I have seen the bird display, and not as described by Mr. Bartlett, although not for one moment doubting the accuracy of such a keen observer, and I am sure I shall be excused for representing only what I have seen, especially as that is sufficiently curious. When I have noticed the head, it has been placed under the right wing; but I should not think this is invariably the case. You are aware that I have previously called attention to the very artistic shading of the large round spots on the secondaries, and my opinion that the bird during display would so place his wings that all the lights on these spots would be upwards or towards the source of light, and the shades downwards, has been confirmed by observation of the living bird."

From my own observation I can fully confirm the statement of Mr. Wood, namely, that the ocelli are so shaded as to represent the light coming from above when the wings are expanded as the bird is displaying itself. In the engraving the ocelli of the secondary feathers nearest the tail have the light side shown nearest the top of the feather, whereas on the first and second secondaries, those which are held nearest the ground and most horizontally, the light is next the shaft of the feather.

The mode in which these ocelli have been produced has been the subject of a very elaborate and ingenious disquisition by Mr. C. Darwin ("Descent of Man," vol. ii., p. 141), to which I would refer those of my readers who desire to enter more deeply into the subject; but the following remarks on the characteristics of the feathers and their employment by the male are so graphic that I need make no apology for quoting them (vol. ii., p. 91):—

"The immensely developed secondary wing feathers, which are confined to the male, are ornamented with a row of from twenty to twenty-three ocelli, each above an inch in diameter. The feathers are also elegantly marked with oblique dark stripes and rows of spots, like those on the skin of a tiger and leopard combined. The ocelli are so beautifully shaded that they stand out like a ball lying loosely within a socket. But when I looked at the specimen in the British Museum, which is mounted with the wings expanded and trailing downwards, I was greatly disappointed, for the ocelli appeared flat or even concave. Mr. Gould, however, soon made the case clear to me, for he had made a drawing of a male whilst he was displaying himself. At such times the long secondary

THE ARGUS PHEASANT DISPLAYING ITS PLUMAGE.

feathers in both wings are vertically erected and expanded, and these, together with the enormously elongated tail feathers, make a grand semi-circular upright fan. Now as soon as the wing feathers are held in this position, and the light shines on them from above, the full effect of the shading comes out, and each ocellus at once resembles the ornament called a ball and socket. These feathers have been shown to several artists, and all have expressed their admiration at the perfect shading.

"The primary wing feathers, which in most gallinaceous birds are uniformly coloured, are in the Argus pheasant not less wonderful objects than the secondary wing feathers; they are of a soft brown tint with numerous dark spots, each of which consists of two or three black dots with a surrounding dark zone. But the chief ornament is a space parallel to the dark blue shaft, which in outline forms a perfect second feather lying within the true feather. This inner part is coloured of a lighter chesnut, and is thickly dotted with minute white points. I have shown this feather to several persons, and many have admired it even more than the ball-and-socket feathers, and have declared that it was more like a work of art than of nature. Now these feathers are quite hidden on all ordinary occasions, but are fully displayed when the long secondary feathers are erected, though in a widely different manner; for they are expanded in front like two little fans or shields, one on each side of the breast near the ground.

"The case of the male Argus pheasant is eminently interesting, because it affords good evidence that the most refined beauty may serve as a charm for the female, and for no other purpose. We must conclude that this is the case, as the primary wing feathers are never displayed, and the ball-and-socket ornaments are not exhibited in full perfection except when the male assumes the attitude of courtship. The Argus pheasant does not possess brilliant colours, so that his success in courtship appears to have

.depended on the great size of his plumes, and on the elaboration of the most elegant patterns. Many will declare that it is utterly incredible that a female bird should be able to appreciate fine shading and exquisite patterns. It is, undoubtedly, a marvellous fact that she should possess this almost human degree of taste, though perhaps she admires the general effect rather than each separate detail. He who thinks that he can safely gauge the discrimination and taste of the lower animals may deny that the female Argus pheasant can appreciate such refined beauty; but he will then be compelled to admit that the extraordinary attitudes assumed by the male during the act of courtship, by which the wonderful beauty of his plumage is fully displayed, are purposeless; and this is a conclusion which I for one will never admit."

The illustration, by the late Mr. T. W. Wood, speaks for itself; its accuracy of detail is remarkable, and I have much pleasure in having been accessory to the publication of the first correct delineation of the display of the Argus pheasant that has been produced.

APPENDIX.

TRANSPORT OF PHEASANTS FROM ABROAD.

ANY PERSONS may be desirous of bringing or sending gallinaceous birds to England, and I cannot therefore do better than reprint the following instructions, which were drawn up for the Zoological Society by Mr. P. L. Sclater and Mr. A. D. Bartlett for the benefit of those desirous of forwarding the various species to England.

"INSTRUCTIONS FOR THE TRANSPORT OF PHEASANTS AND OTHER GALLINACEOUS BIRDS.

"1. For exportation, birds bred or reared in captivity should, if possible, be procured. But if this cannot be done, the following rules should be attended to as regards wild-caught birds:

"2. As soon as the birds are captured, the feathers of one wing and of the tail should be cut off tolerably close to their bases. The birds should be placed in a room lighted only from a skylight above, and having the floor sprinkled with gravel or sand, mixed with tufts of grass and roots and a little earth. Among these the food should be thrown. A tame bird placed with the wild ones is of great advantage, because this bird will induce the new captives to feed. The birds should be kept in this way until they have become tame and are fit to be transferred to the packing-cases.

"3. The food should consist of grain and seeds of various

kinds, berries, fruit, insects, green food (such as cabbage, lettuce, &c.), bread or soaked biscuit, chopped meat, boiled eggs, &c.

" 4. Travelling cages are most conveniently made of an oblong shape, divided into compartments about eighteen inches square, and not higher than just sufficient to allow the birds to stand upright in them. They should be boarded all round, except in front, where strong wire netting may be employed—although, if the birds are at all wild, wooden bars, close enough to prevent the inmates from escaping between them, are preferable.

" 5. Every compartment should have the top on the inside padded with canvas, as, if this is not done, the birds are very liable to injure their heads by jumping upwards.

" 6. A movable feeding-trough should be fixed along the front of each compartment; one-third of this should be lined with tin, pitch, or otherwise made to hold water; the remaining two-thirds will hold the food.

" 7. Coarse sand or gravel should be kept strewn on the bottom of the cages, and a supply of this should be sent along with the birds, as it is necessary to them for the healthy digestion of their food.

" 8. The front of the cage should have a piece of coarse canvas to let down as a blind to keep the birds quiet; and, in order to give them air, round holes should be bored at the back of the box in the upper part.

" 9. The box should be cleaned out when the birds are fed, through the opening in front made by removing the feeding trough, care being taken that this opening is not wide enough to let the birds escape.

" 10. In order to supply the birds with green food during the voyage, a few small trays (the same as are used to hold the sand or gravel) may be sown with seeds, such as rape, mustard, or any quick-growing vegetable. The green food thus produced should be cut for them from time to time, and the sand and roots afterwards thrown into the cages."

TRANSPORT OF PHEASANTS FROM ABROAD. 229

For securing any recently-caught or very wild bird in such a manner that it is unable to injure itself by dashing against the sides or top of the cage, the plan used by falconers, and termed by them brailing, is advantageous.

To secure each wing, two pieces of string or tape of equal length must be taken, and two knots tied, as shown in

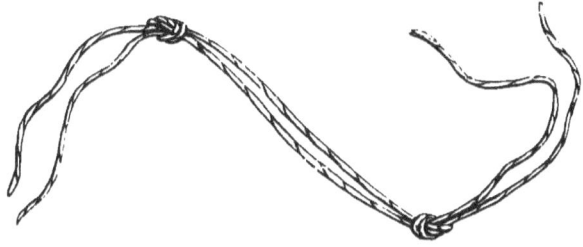

FIG. 1.

Fig. 1, so as to form a central loop with loose ends. This loop must be of a size proportionate to that of the wing of the bird to be secured. When used the loop is passed over the fore part of the wing, and one set of loose ends are brought up behind, between the wing and the body, and

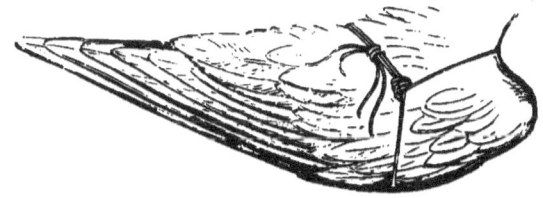

FIG. 2.

secured by being tied to the other set, as shown in the lower figure. If this is properly done, there will be no pressure on any part of the wing, nor need a single feather be ruffled or deranged; nevertheless flight is entirely prevented, as the bird has no power of expanding the wing. When properly brailed the wildest bird may be placed on the ground, where it can run about freely, but without the least power of flight. This plan is one of great utility in the transport of very wild birds, as they are quite unable to dash themselves against the roof or sides of the cage in which they are inclosed.

I need hardly say that should a bird be confined a long time in this manner it would be necessary to loosen the wings alternately, otherwise a stiff or contracted joint might ensue. This would be obviated by allowing the bird the free use of each wing for a short period.

INDEX.

Acorns injurious in confinement ... ...*page* 82
Alarm guns 63
America, North, the pheasant in 37
Amherst pheasant... 199
Argus pheasant 220

Barnes, Mr. J., on feeding in coverts 54
Barren hens 60
Bartlett, Mr., on rearing young pheasants 113
Bartlett, Mr., on Sœmmerring's pheasant 168
Bartlett, Mr., on the transport of pheasants 227
Baskets for transporting pheasants 94
Beakless pheasant 76
Bennett, Dr., on Reeves's pheasant 177
Black-throated Golden pheasant 190
Blindness in young pheasants 132
Blyth on call-note of Reeves's *181
Bohemian pheasants 149
Bones, crushed, use of 88
Brailing hens with young 118

Carbolic acid for "Gapes" 130
Carreau, Mons., on the Amherst pheasant 201
Carr-Ellison on formation of coverts 41
Cats destructive to pheasants 75
Chinese pheasants 155
Cobbold, Dr. Spencer, on "Gapes" 126
Cock pheasants sitting 16

INDEX.

Cocks, proportion desirable ... ...page 60
Common pheasant wild in Albania ... 38
Coops for young pheasants ... 116
Cordeaux, Mr., on power of flight ... 10
Corsica, wild pheasant in ... 38
Cost of rearing ... 123
Courtship, display of plumage during... 12
Coverts, formation of ... 41
Coverts, food in ... 46
Crane-fly grubs eaten by pheasants ... 6
Crows destructive to young pheasants... 69
Cramp in young pheasants ... 131
Crook's arrangement for pheasantries ... 81
Cross-bred pheasants in coverts ... 166
Crossoptilon mantchuricum ... 212
Crossoptilon tibetanum ... 212
Crowing ... 12
Curd as food for young pheasants ... 111
Custard as food for young pheasants ... 111

Darwin, Captain, on mock pheasants ... 63
Darwin on cross-bred Sœmmerring ... 174
Darwin on display of Argus pheasant... 224
Dawkins, Mr. W. B., on introduction into England ... 25
Digestive organs ... 8
Diseases of pheasants ... 125
Distribution throughout Europe ... 35
Distribution throughout Great Britain ... 32
Domestication, pheasants not capable of ... 21
Douglas, J., on rearing young pheasants ... 114

Eared pheasant ... 212
Egg-eating by pheasants ... 83, 96
Egg testers ... 106
Eggs, purchasing stolen ... 107
Elliot, Mr. D. G., on hybrid Amherst... 203
Elliot's *Phasianidæ*... 1
Enteritis in pheasants ... 135

INDEX.

Euplocamus nycthemerus ...page	206
Exportation of pheasants	227
Feeding in coverts ...	51
Feeding-troughs objectionable	53
Firs, species best adapted for coverts ...	42
Flight of pheasants	10
Food of pheasants ...	3
Food of pheasants during transport	227
Food for young	111
Foxes, driving them from vicinity of nests...	74
Gapes in pheasants...	126
Golden pheasant	188
Golden pheasant wild in Oregon ...	37
Gentles as food	112
Gentles from seaweed	112
Gould on *Phasianus sœmmerringii*	174
Gould on *Phasianus torquatus*	156
Gould on *Phasianus versicolor*	162
Grasshoppers eaten by pheasants ...	49
Greece, distribution of pheasants in	39
Gurney on Japanese pheasants	167
Harting, Mr., on pheasant in Middle Ages...	27
Harting, Mr., on rooks destroying eggs	70
Hatching in confinement	104
Hedgehogs destructive to eggs	75
Heine, Mr., on habits of Japanese pheasant	163
Heine, Mr., on Sœmmerring's pheasant	171
Hens, varieties best adapted for hatching ...	103
Hewitt on Golden pheasants...	194
Horne on Reeves's pheasant ...	183
Hybrid Reeves's pheasant	187
Impeyan pheasant ...	215
Introduction of pheasants into England	25
Introduction into Ireland, date of	31

INDEX.

Introduction into Scotland, date of ...*page*	31
Introduction into St. Helena	159
Introduction into New Zealand	35
Introduction into Samoa	36
Introduction into North America...	36
Introduction into Oregon	37
Japanese pheasant	162
Jerusalem artichokes for pheasants	57
Jeffries on pheasant rearing	99
Jess for tethering hens	118
Kaleege	210
Kestrel occasionally destructive to young pheasants	72
Klein, Dr., on diseases of pheasants	131
Laying, date of	17
Latham, Dr., on Reeves's pheasant	178
Leno, Mr., on rooks destroying eggs	65
Leno, Mr., on pens for pheasantries	79
Lettuce, use of, for young birds	84
Lilford, Lord, on Reeves's pheasant	181
Lilford, Lord, on the introduction by the Romans	25
Lort, W., on feeding in coverts	54
Macgillivray, description of the common pheasant	146
Macgillivray on food of pheasant	4
Maggots from seaweed	112
Male plumage, assumption of, by female	138
Mantchurian Eared pheasant	129
Marco Polo on Reeves's pheasant...	177
Mayes, Mr. J., on Reeves's pheasant	186
Millais, Mr. J. G., on Reeves's pheasant	184
Mock pheasants, to make	62
Monaul	215
Moorhen destructive to young pheasants	71
Naumann on the pheasant	2, 3
Nesting	13

INDEX.

Nests in trees ... *page*	15
Net for catching pheasants in aviaries	87
New Zealand, successful acclimatisation in	35
Non-domesticity of common pheasant...	21, 24
Oak-spangles as food for pheasants	5
Ogilvie-Grant, Mr., on *P. scintillans*	176
Open pens for pheasants	84
Orpington disease in pheasants	135
Partridges laying in pheasants' nests	14
Pens for pheasants	78
Perry, Commodore, on Sœmmerring's pheasant	170
Phasianus chrysomelas	144
Phasianus colchicus	143, 166
Phasianus decollatus	160
Phasianus elegans	205
Phasianus insignis	144, 153, 160
Phasianus mongolicus	153, 160
Phasianus persicus	144
Phasianus pictus	188
Phasianus principalis	152
Phasianus reevesii	178
Phasianus scintillans	175
Phasianus shawii	141, 160
Phasianus sœmmerringii	169
Phasianus strauchi	144
Phasianus superbus	179
Phasianus torquatus	155, 166
Phasianus veneratus	178
Phasianus versicolor	162
Phasianus vlangali	144
Phasianus wallichii	187
Pied pheasants	151
Pinioning young birds	193
Potatoes, boiled, use of	82
Prince of Wales's pheasant	152
Raisins for pheasants	57
Rearing in preserves	58

INDEX.

Rearing young pheasants	page 58, 109
Reeves's pheasant	178
Ring-necked pheasant	157
Rooks destructive to pheasants	65
Roup in pheasants	126
Saurin on Reeves's pheasant in Pekin	180
Scent, suppression of, during nesting	73
Sclater, Mr. P. L., on transport of pheasants	227
Sexes, due proportion required	59
Shot, pheasants poisoned by eating	141
Silver pheasant	206
Sinclaire, Mr., on Golden pheasants	190
Sitting hens, arrangement recommended	104
Skin disease	133
Slow-worms eaten by pheasants	7
Sœmmerring's pheasant	269
St. Helena, pheasants in	159
Stevenson on cross-bred pheasants	167
Stevenson on pheasants	158, 167
Stone on the introduction of Reeves's pheasant	179, 181
Swimming, examples of	11
Swinhoe, Consul, on the Silver pheasant	206
Swinhoe, Consul, on the Eared pheasant	213
Tameness, examples of	22
Tethering hens with young	117
Thaumalea amherstiæ	199
Thaumalea obscura	190
Thaumalea picta	188
Theobald, Mr., on gapeworm	128
Thompson on food of pheasant	4
Transport of pheasants, instructions for	227
Turkey-hens as rearers	108
Vegetable food, necessity for	84
Vipers devoured by pheasant	7
Wallace, Mr., on the Argus pheasant	220
Water, catchpools for, in coverts	56

Water, rearing young without ... ...*page*	119
Waterton on non-domesticity of pheasant...	24
Waterton on formation of coverts	48
Weight of common species	19
White pheasants	150
Whyte, Col. J., rooks destroying eggs...	68
Windows broken by pheasants	9
Wireworms eaten by pheasants	6
Wood on the display of Gold pheasant	189
Wood on the display of Argus pheasant	223
Wool, death of young pheasants caused by	139
Yew-leaves poisonous to pheasants	139

www.ingramcontent.com/pod-product-compliance
Lightning Source LLC
Chambersburg PA
CBHW031933230426
43672CB00010B/1914